THE
REGENERATIVE BRAKING
STORY

ISBN 1 905304 161
ISBN 9781 905304 165

Title Page illustration: Each OK-45 B controller top within Glasgow's fleet of Regen trams was fitted with this plate to remind motormen that their driving technique should be appropriate to this type of tram.

Design and Computer Origination: John A Senior

Trade Distribution and Sales Enquiries
MDS Booksales 128, Pikes Lane Glossop Derbyshire
01457 861508

or

Scottish Tramway & Transport Society
PO Box 7342 Glasgow G51 4YQ

Website stts.glasgow@virgin.net

THE REGENERATIVE BRAKING STORY

BY

STRUAN JNO T ROBERTSON
AND
JOHN D MARKHAM

PUBLISHED JOINTLY BY

THE SCOTTISH TRAMWAY & TRANSPORT SOCIETY
AND
VENTURE PUBLICATIONS LTD

INTRODUCTION

Recognising that the topic of Regenerative Braking tramcars had never really been given much attention barring some articles many years ago in *Modern Tramway* and *Tramway Review*, I suggested to Struan Robertson that he might like to pen a few words for The Scottish Tramway & Transport Society's annual magazine *Scottish Transport*. I also suggested that he might like to major on the Glasgow fleet. The city had the largest of the 1930s' fleets of 'Regen' trams in Britain and one of which he had first hand personal knowledge and experience.

I expected something extending to around nine or ten pages that could be relatively easily illustrated. I did *not* expect around ninety pages covering the whole topic, not just the Glasgow trams!

However, Struan Robertson is a modest man and was very conscious of his own perceived limitations. "I am a retired G.P.," he protested, "and I would really like some validation of my text".

Struan Robertson

A chance, but fortunate, encounter with John Markham at Crich in the course of the fortieth anniversary of the closure of the Glasgow tramway system confirmed that he would assist with the electrical technicalities. Unknown to me at the time, Struan had been encouraged to make contact with John Markham by his fellow historians. John is probably the foremost and most knowledgeable dc traction engineer we have and his offer was welcomed with open arms. Although he probably did not know what he was letting himself in for at the time, his assistance has proved to be so invaluable that he soon became a co-author, bringing his own knowledge, experience and authority to a previously neglected, and sometimes complicated subject.

With one author in Dornoch and the other in Stockport, and few meetings across a table, the editor has had to act as go-between and co-ordinator. Each author writes in his own style and I decided early on not to try to merge the two together so that 'you can't see the join'. In the end, each has selected chapters within his own particular sphere of knowledge although some minor overlapping has occurred.

The story that emerges is one of an idea that was ahead of its time or relied on technology that was unable to cope with it. It had to harness primitive tramway equipment, long before the availability of solid-state electronic technology that makes regenerative braking commonplace and beneficial in the hi-tech trams of today. It is a matter of some regret that there are so few examples in Britain to reap this benefit.

John Markham

Having given the matter considerable thought, the book has been divided into three distinct sections. These comprise, firstly, the Historical Chapters, followed by the Technical Chapters and, finally, the Appendices where some explanatory and supplementary detail has been included that might otherwise have detracted from the flow of the historical text. While not devaluing in any way the technical content, those with an interest in the historical narrative will find that this stands on its own. They are commended, however, to read the technical explanations that confirm why history sometimes followed the path that it did.

Ian G McM Stewart
Editor
Newton Mearns
January 2006.

CHRONOLOGY

1846	Birth of John Smith Raworth
1866	Samuel Alfred Varley invents the self-exciting dynamo
1892	Varley is declared inventor of compound winding
1902	Johnson-Lundell experiments with regenerative equipment in Newcastle
1903	Prototype Raworth demi-car enters service in Southport
1904	Raworth Traction Patents Company registered
1905	Run-away collision in Halifax involves two demi-cars
1906	Birmingham Corporation bans regenerative cars from its tracks
1907/8	Johnson Lundell Company fights Raworth Traction Patents in court
1909	Demonstration trolleybus built by RET Company
1909	Last demi-car built (for Maidstone)
1910	Leeds and Bradford Corporations obtain parliamentary powers to operate trolleybuses
1911	Rawtenstall collision results in Board of Trade Report vetoing regenerative trams
1912	Raworth Traction Patents Company wound up.
1926	First double-deck regenerative trolleybus placed in service (Wolverhampton 33)
1930	Paris Conference of International Tramway & Light Railway Union
1930	Royal Commission Report recommends replacement of tramways
1931	Manchester tram 420 is equipped with regenerative equipment
1932	Equipment from Manchester 420 passed to LCCT
1932/33	Glasgow Corporation experiments with regenerative equipment with car 305
1933	First experiments with regenerative equipment on Edinburgh tramcars
1934	Leeds Corporation prototype Middleton Bogie 255 enters service with regen equipment
1933	Extension of experiments with regenerative equipment in London
1933	Regenerative equipment returned to Manchester 420 for further tests
1934	Halifax experimental regenerative car enters service
1934/35	Production batch of regenerative cars enters service in Edinburgh
1934/35	Glasgow's production batch of 40 regenerative cars enters service
1935	Experimental regenerative car enters service in Johannesburg
1936	Birmingham Corporation persuaded to experiment with regen equipment in car 820
1937	Johannesburg places in service 50 production streamliner regenerative cars
1938	Glasgow experiments with field control
1962	Last UK traditional city tramway closes in Glasgow
1992	Manchester Metrolink system opens using regenerative braking.

Acknowledgements

I n my youthful, escalating, interest in the tramway world my understanding was markedly broadened and enhanced by reading the explanations of the technology written by 'Eltee' – the nom-de-plume of the late BJ Prigmore, in the early numbers of *The Modern Tramway*. His writings were concise, and his descriptions accurate and readily understood. He opened the door to a much less superficial understanding of a subject difficult to pursue due to the lack of published detail in those days. In writing this book I should hope at least to emulate, though not to equate with Eltee's lucidity, and to pass on to others some of the work he may not have covered. John Markham has been my mentor, in later years.

Nonetheless, I have launched into this work with no mean degree of trepidation, following in the footsteps of the eminent predecessor in my profession – Dr HA Whitcombe – the only medical doctor ever to have read a paper on tramcars (admittedly steam ones) before that august body of Railway Engineers, 'The Institution of Locomotive Engineers' whose President at the time was Mr WA Stanier, Chief Mechanical Engineer of the then London, Midland & Scottish Railway, and of Great Western Railway training.

There is a great paucity of information, available mainly in journal form, of the story and technical detail of regenerative braking, other than in the sporadic articles and notices in the various trade periodicals of the time – difficult to access in the far north of Scotland – but God Bless the wonderful inter-library loan system for quite a lot of material.

As in so many books on tramcars, there were so very many persons who selflessly contributed so much detail that a tally is invidious and impossible. Nevertheless, to one and all who have contributed to this book in kind or advice; in guidance or flat correction, I thank you all sincerely.

Gracious and unreserved acquiescence to the use of fellow authors' works, and notes, has never been turned down, and outstandingly we thank Ian Yearsley for the bulk of the demi-car material, without whose diligent and intensive research this book would never have been complete. An approach to him to join in co-authorship was declined owing to his overwhelming teaching and journalistic commitments.

Charles C. Hall of Sheffield, I met in the Navy, and had the temerity to contradict in the Ward-Room over some Glasgow detail – only to be firmly put right from his infinitely more vast historical knowledge. He became my closest friend thenceforward and taught me so much over the years about transport, its technology and history, and goaded me on to writing about the subject. My deepest thanks go to him for the constant loan of books, notes and extracts from his own life-long study of this fascinating topic.

To Philip Groves, also, my very real thanks for his encouragement and likewise unlimited access to his wonderful writings on Manchester and Birmingham, not to mention those most interesting letters from the Metro-Vick Company on the assessment and vicissitudes of the Manchester experimental regenerative car that shed much light on the history of that episode.

To Philip's friend and co-author of *City of Birmingham Tramways Company Limited* in *Tramway Review* 1992, Mr Peter Jacques, my most grateful thanks indeed for contributions of Birmingham interest, otherwise unobtainable.

To George T McLean – for great encouragement and the loan of his library of electric textbooks

To Messrs. Alan W. Brotchie, Ian L Cormack and Ian A. Souter, in the same category of transport authors, have my thanks for detail and photographs of regenerative history, and cars that otherwise had not come my way. In the same way Messrs. Brian Longworth, Hugh McAulay and Geoff Price have shown unremitting generosity in complementing information and correcting false notions!

To Henry Campbell, Electrical Department Foreman, of Coplawhill Car Works – and his many colleagues – for befriending and explaining everything in the Glasgow Car Works, access to which I have to thank ASE Browning, one time Curator of Technology of the Glasgow Museums & Art Galleries, for that privilege.

Then Crich! – the wonderful National Tramway Museum, where every member would turn round and offer immediate help at the drop of a hat! Kind mention must go to the Reverend David Tudor – now retired Principal Driving Instructor and Examiner to the National Tramway Museum, for the chapter on Continental Braking methods, an area I am more than weak on. To the gracious, imperturbable and truly kind Rosy Thacker, then Librarian to the National Tramway Museum for immense help from the research aspect, not to mention their then Photographic Curator, Glynn Wilton, whose hands were tied by building works from access to his records, yet managed to provide some of the excellent illustrations herein. To all of them, my unbounded thanks.

Then to everyone involved in the production of this work: the Scottish Tramway & Transport Society for agreeing its undertaking: our Editor, Ian G McM Stewart who kept certain noses to the grindstone and was utterly helpful throughout; also Alistair J Douglas of Strathaven, specialist in Transport Photography, for amazing patience with a demanding customer, and consummate skill in the accomplishment of rejuvenation of Brownie Box negatives of 60 years duration; John Senior for making available photographs from the Senior Transport Archive (STA) and his enthusiasm for the project; David Thomson for overseeing the completion.

Finally, at the advice of Philip Groves and Charles Hall, both bade me get in touch with John D. Markham – recognised as probably the most outstanding Traction Electrical Engineer in the country. To my very great delight, John agreed to co-authorship. His succinct, clear, interesting and very readable contributions have rendered this book utterly authentic and, what is more, will have made it useful to a wider interest than originally contemplated. Thank you, indeed, John, for your patience, as well as for your delightfully interesting contributions.

John has added his own acknowledgements: In the process of writing these essays for this book, I have had the pleasure and advantage of help and advice from several of my friends and acquaintances whom I should like to thank, and have remembered formally. First of all the staff at AEI Archives, and Geoffrey Claydon for great personal help, Danny Cohen in America, Peter Deegan of Rawtenstall, Edward Oakley of London, the late Percy Lawson, Alan Ralphs – my colleague in Trafford Park, and Jim Soper, author of the definitive works on Leeds Transport. To all go my grateful thanks for help, advice and necessary correction.

In conclusion and of supreme outstanding importance, to my very darling and most brave late wife, Rhoda, who suffered recurrent surgical operations and the most distasteful medications while much of this work was proceeding (one of the reasons why it has been so very long to mature) God Bless, and thank you for your forbearance, understanding and wonderful encouragement.

Struan Jno T Robertson
Dornoch
January 2006

CONTENTS

CONTENTS

GLOSSARY OF TERMS AND ABBREVIATIONS

ac	Alternating current
AEB	Automatic emergency brake
AEC	Associated Equipment Company
AEI	Associated Electrical Industries Limited
Back-emf	Back electro-motive force
BMT	Birmingham & Midland Tramways
BT-H	British Thomson-Houston Company Limited
BO	Blow out coil
dc	Direct current
EMB	Electro Mechanical Brake Company Limited
emf	Electro-motive force
GCT	Glasgow Corporation Transport (or Tramways)
GEC	General Electric Company Limited (the British Company)
GTO	Glasgow Tramway & Omnibus Company Limited
hp	horsepower
Hz	Hertz (frequency in cycles per second)
LCC	London County Council
LUT	London United Tramways
M&T	Maley & Taunton Limited
MAR	Mercury Arc Rectifiers
MET	Metropolitan Electric Tramways
mph	Miles per hour
Met Cam	Metropolitan Cammell Limited
Metro-Vic	Metropolitan-Vickers Electrical Company Limited
M-V	Metropolitan-Vickers Electrical Company Limited
MTTA	Municipal Tramways and Transport Association
PWD	Permanent Way Department
Regen	Regenerative
rpm	Revolutions per minute
RTP	Raworth Traction Patents Company Limited

PART ONE
HISTORICAL CHAPTERS

SETTING THE SCENE

The technology that brought about Regenerative Braking is set out in the technical Chapters 10-13. Here, the scene is set for the historical chapters that will take the reader through the first and second phases of regenerative braking in Britain, the near-demise of the traditional tram culminating in the slow and painful birth of the new generation trams that benefit from continental practice.

You would not have lifted this book to read had you not already a good general interest and knowledge of tramways. Let us take it a stage further into the realms of 'how it works' thus significantly extending a basic appreciation of, and pleasure in, your personal interest. All electric tramcars may be divided into two main phases of propulsion, separated from each other in this country by roughly two decades of total sterility – the 1960s and 1970s – when practically no towns or cities had electric tramways at all. They had all been scrapped! Blackpool and the Isle of Man were, of course, the exceptions that proved the rule. These two main phases, divided by the acme of supremacy of the internal combustion vehicle were, firstly, the simple series-parallel tramcar of the first sixty odd years straddling the late nineteenth and twentieth centuries, and the solid-state electronically controlled tram-trains of the latter quarter of the twentieth century onwards.

There is, of course, regenerative braking control which itself may be divided into three disparate phases within the above general classification. That of primary regeneration, of the 1902-1911 phase – almost entirely of John Smith Raworth's period – was followed by the 1930s' phase of resurgence of interest in the more modern regenerative equipments, and the very much more up-to-the-minute phase of solid-state electronics that emerged in the late 1970s. This latter, third, phase, is unique in discarding all previous design details including double-decking. This is now a thing of the past and there is now established the continental single-decked, low-loading, articulated tram-train along with a swathe of ultra-modern equipment that shall be left to subsequent writers to explore once this avalanche of current invention (now with regenerative braking as standard) has bedded down. Extremely slow, and desultory in its introduction in this country, the third phase spread very gradually at home, largely against political interest which was vacuous on the subject, and, on account of expense, due to Britain's policy of ridding itself of most of its vehicle builders, not to mention steel works for rolling tramway section rails. It is these first two classifications with which this book very largely confines itself in an attempt to cover historically and technically what has gone before, while paying tacit recognition to the tremendous advances going on in Phase III, still in its elementary stages!

Regenerative braking has been chosen as having scant coverage previously and there is, already, a very marked paucity of available information. What was available was gathered with the selfless help of a great many from both within and outwith the official Tramway world. The picture of these first two Phases of Regeneration has been recreated and presented so that it should not recede into the limbo of forgetfulness. Businessmen often, and construction engineers occasionally, have been known to reject and forget the history of the origin of systems as out-of-date inconsequence, and sally forth into an exhibition of modern concept devoid of reference to what has been found necessary before. The failure to use water-drainage rails in new construction jumps to mind!

Interestingly the first general phase of electric tramways in this country was epitomised by open-topped, double-decked and double-ended four-wheeled tramcars. These had short-circuit or shunt transmission with unventilated motors devoid of interpoles or roller bearings. They were of slow-speed, accomplishing generally less than 8-10 mph (largely due to street congestion). This progressed in natural steps ultimately to total enclosure of the vehicles and their fitting with high-speed motors, then capable of 28-30 mph, many with equal-wheel bogies with all axles motored. On the Continent, however, the standard had nearly always been of single-deckers (Paris being one of several exceptions operating double deckers) either of four-wheeled or bogie types, with one or more trailer coaches. Very often these had single-ended drive with turning circles at peripheral termini which made for a quicker turn-around.

As mentioned, the 1960s saw all but a couple of UK tramway systems – not to mention trolley-bus systems within the decade – yielding to the supremacy of the cheaper to build, and run, ubiquitous access motor-buses. Quite soon, with approved lengths progressively increasing, and their taking up more road space, these succumbed to increasing traffic congestion and the commuters' desire for personal transport. By the 1990s three or four British cities were turning to the use of modern continental-type single-decker tram-trains to cope with demands. In some areas this has been with very marked success.

This has been the much-compressed overall picture of the development of the tramway systems of this country from which we are now going to select but one, very remarkable, but elusive feature of its growth industry: regenerative braking.

Regenerative braking for tramcars was first introduced in the Edwardian era when travel usually involved elemental open-top trams like this Newcastle example. *(Photo: BJ Cross collection)*

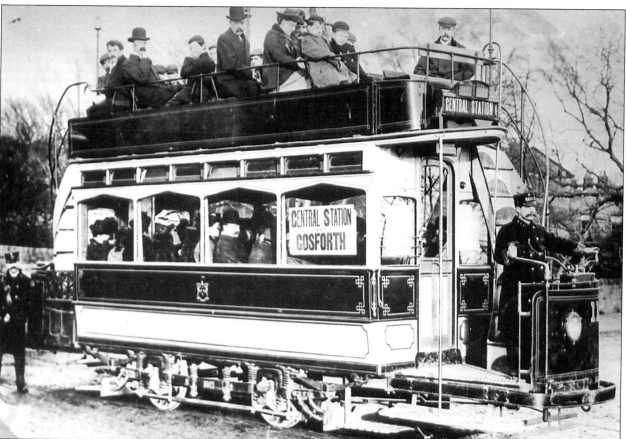

THE FIRST PHASE OF REGENERATIVE BRAKING – AND THE DEMI-CARS

The history of regenerative braking in tramcars is one of a long, slowly maturing process, set into three very definite and disparate phases over a period of at least 70 years, if not longer. The first phase – that of the pre-First World War era – while essentially a feeling-of-the-way process – was the work and invention of practically one man alone. This was John Smith Raworth. Struan Robertson takes us chronologically through the development of the equipment and its particular application in demi-cars.

BACK TO THE BEGINNING:

In September 1866, the first self-exciting dynamo – which made electric traction a practical possibility – was invented and constructed by Samuel Alfred Varley. For this invention he was awarded the Gold Medal of the 1885 International Exhibition, 19 years after it was made. He followed this up with his second important advance in dynamo construction: *Compound Winding* from which it was but a short step in principle, although a longer step in time, to the compound-wound regenerative motors of the early 1900s.

Interestingly, the firm of Brush of America took out an elaborate patent in 1878; ten months after the specifications of Varley's "Compound Winding" patent had been published. The Brush description was virtually in the same wording as Varley's 1876 patent!

A Scots firm, in consultation with Varley, contested the American Brush Corporation's action on commercial grounds in the Scottish Courts against the Brush Corporation, for damaging their business by threatening their customers. This strategy immediately changed the Brush Corporation's legal position of the assailant to that of defender, and they were thereby compelled to fight on ground chosen by the Scottish firm, wherein English Counsel were not allowed to plead. The combination of England's ablest patent barristers and scientific experts was thus broken. They had, until then, carried all before them in the English Law Courts and the Anglo-Brush Corporation had almost entirely relied upon them. Every effort was made by the Brush Corporation to have the case removed to the English Law Courts without success. When the hearing of the case for the reduction of the Brush patent was about to start, that firm approached the Scottish firm offering such very favourable terms as to induce them to compromise that their legal advisers advised their acceptance. Varley protested against any compromise as a breach of good faith with him and the Scottish firm offered him the only alternative – the raising of a guarantee fund. Such a guarantee fund was raised and the case was heard in the Court of Session, Edinburgh, in November 1888. This battle was won by Varley and later, when the case went to appeal in the Inner House of the Court of Session, Brush's defeat was converted into so complete a rout that the wealthy Anglo-Brush Corporation found it immediately necessary to go into liquidation at once, and to

Brush's defeat was converted into so complete a rout that the wealthy Anglo-Brush Corporation found it immediately necessary to go into liquidation at once

reconstitute themselves as a new company. Lord McLaren, who delivered judgement in the Inner House stated:

> *"I have difficulty in understanding how it is that a considerable number of able, and distinguished, men should have been persuaded to give their evidence as to the alleged insufficiency of Varley's as an anticipation…and indeed I do not see how the validity of the Brush patent can be maintained, because the two descriptions are practically identical … I can find nothing of substance in Brush that is not in Varley, and I cannot help adding that if there be any difference, Varley's description is the more easily understood of the two."*

The Brush Corporation took the case to the House of Lords and although the two ablest of English patent Counsel argued the matter for nearly a week on behalf of Brush, the Lords did not consider it necessary to hear the other side and later, on April 5th, 1892, in a series of elaborate judgements delivered by the Lord Chancellor and the late Lords Herschell and Watson, in which the other Lords expressed their concurrence, they unanimously confirmed the two previous judgements of the Scottish Courts and declared Varley to be the inventor of Compound Winding.

[The Tramway & Railway World: Vol.III: 1903, 8th October : 'The Career of Mr Samuel Alfred Varley – the Inventor of Compound Winding' : pp256-260]

That Varley did not invent Regenerative Braking, *per se,* but did invent the compound-wound dynamo, is of great moment in the history of the former, for John Smith Raworth happened to be a very senior electrical engineer in the firm of the Brush Corporation – later the Brush Electrical Engineering Company Limited – of Loughborough, and undoubtedly exploited the short step from his firm's legal shenanigans to the compound-wound regenerative braking motor of 1902-03, on his demi-cars. Varley, meanwhile, had made his name in trans-Atlantic telegraphy as well as perfecting the dynamo – perhaps the most striking invention of the 19th century. He was certainly a genius far ahead of his time.

The use of regenerative braking was designed to save on current. Another method adopted in London and Newcastle was to use trailers where increased capacity could also be provided. Here are two Newcastle trams being so operated and both numbered '146'.

(Photo SITA collection)

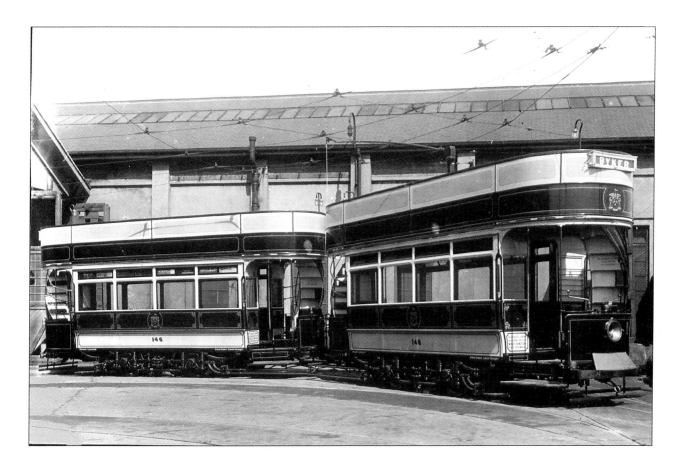

An Overview:

It was not as if there was a dearth of other transport engineers researching the concept at the time. It was simply that none of them was able to make regenerative braking work except for one firm of electrical machinery makers in London – the Johnson-Lundell Company. As will be seen in the technical chapters, that firm did invent a different approach to the harnessing of the regenerative capabilities of the dc traction motor and did reach the stage, in 1902, of having at least one tramcar fitted up so that extensive trials were held in Newcastle-upon-Tyne.[1] Trammelled by its, as yet, unproven value and the infancy of scientific circuitry involved in effecting regeneration, the concept did not survive its first decade in this country. John Raworth was an electrical engineer with deep insight into its transport aspect. He was pre-eminently of genius stature in his understanding of what was required but hamstrung practically for the lack of advancement in motor design, then without adequate ventilation and interpoles. All this was to come, but in the Second Phase of Regenerative Braking. Even then, it was not enough!

It was because of its immediately demonstrated amazing economy in electric power, constituting its outstanding attribute, that it cornered a certain new, and previously non-existent, market in small tramcars. Raworth had conceived this and built to exploit his invention. He appeared to be making the grade financially had it not been for the 'Green eye of the Little Yellow God' of jealousy of the Johnson-Lundell firm who thought they might make a thing or two out of having been some six or nine months ahead of Raworth in exhibiting a regenerative tramcar. They sued him for a breach of patent. The legal profession made the most they could of this through procrastination and the upshot was that the Johnson-Lundell firm had to mortgage their works to pay the legal fees. Raworth lost a deal of trade by this and, when a series of accidents culminated in the most severe one at Rawtenstall in 1911, he voluntarily wound up his firm and retired from competition. His little tramcars, which were called 'Demi-cars', battled on for another decade, or more, before fading out of the picture. Meanwhile all the regenerative equipments were withdrawn on the recommendation of the Board of Trade and this terminated the first phase of regenerative braking in tramcars. Nearly twenty years were to elapse before the first resurrection of the system, as will be seen! The Second Phase of Regenerative Braking in Tramcars really commenced in 1930 in this country, although the following chapter will show that it had started some five years before in trolleybuses. Virtually prohibited by the Board of Trade following the Rawtenstall accident, the tramway movement was more than extremely reticent to so much as consider the use of the system despite being perfectly aware of the desirability to do so. Rates for electricity were soaring in the post-World War I depression. Indeed practical research had never quite stopped in the non-railed road traffic world; it was too valuable a commodity and for quite a while electric railways had revelled in this economic value. The Norfolk & Western Railway in USA ran electric locomotives with 3-phase motors off single phase, 11kV at 25Hz overhead using regenerative braking from 1915.[2] So it was not unnatural that individual tramway engineers had kept the concept at the back of their minds.

The trolleybus interest in regenerative braking in this country had gathered momentum in the mid-1920s. Its use became increasingly widespread and valuable in that field of transport from then on and must have stirred the minds of many tramway engineers! A combined Franco-German tramways interest blossomed forth with considerable success on the continent. This, coupled with a compelling demonstration of successful tramways application by the Paris Tramways, finally kindled the light of battle against the Board of Trade veto and resulted in the Second Phase of Regenerative Braking in Tramcars in Britain.

Despite the advances in motor design and technology during the intervening years, it was – sadly – yet too early for the genuine value of regeneration to be demonstrated. There were still too many drawbacks that had not been worked out. Anyway, World War II put an end to all such research and experimentation. Lack of money and the demise of electric street transport following the War once again annihilated all interest in the subject. However, regardless of the out-and-out political triumph of the internal combustion engine, a new field of electrical interest was undergoing development; that of solid-state electronics. Three factors have brokered the eventual full-blooming success in the 1980s of regenerative braking in tramcars: the glut of total motorbus command of the streets and roads together with the de-restriction of bus services and the leadership of the state-of-the-art solid-state electronics in continental street tramways,[3] – the Third Phase! This will be touched upon later.

The legal profession made the most they could of this through procrastination

THE JOHNSON-LUNDELL SYSTEM.

At the invitation of the directors of the Johnson-Lundell Electric Traction Company, a party of electrical engineers and technical Press representatives paid a visit on Thursday to the works of the company at Southall to witness a demonstration of their regenerative system. The exhibition took place upon a circular track laid upon land adjacent to the works. A tramway truck had been equipped with the apparatus, and this was operated by means of an overhead trolly wire. To give the truck the approximate weight of a loaded tramcar, four spare motors are mounted upon the platform; these also served the purpose of showing the difference in the design of the regenerative motor, as compared with the standard series parallel type. Mr. Johnson received the company, and gave a brief description of the system. The visitors then mounted the platform of the truck, and were afterwards taken at various speeds round the track, Mr. Johnson meanwhile further explaining the various details of the working of the apparatus. Briefly put, the system is a combination of the series parallel and compound systems, the use of each being determined by the speed at which the motors are driven and their retardation for generative purposes. The transition of field characteristic from simple series for propelling, to compound regeneration for retarding, and vice-versa, effected through the medium of a third controller, which automatically responds to thumb pressure acting upon a button fixed to the top of the handle of the platform controller. With the exception of a starting résistance, there is a complete abolition of armature circuit resistances, a saving of 15 per cent. in gross input of current being claimed under this head. A feature of this system is the fact that any failure of the controller regulating the characteristic of the motor would not cause the withdrawal of the car from service but would leave it with an efficiently operative series equipment.

A contemporary description reproduced from the *Tramway & Railway World* of 1905. *(STA)*

PHASE I:
REGENERATIVE BRAKING IN TRAMCARS – IN BRITAIN:

The Tramways Act, 1870: *33&34 Vict. Chapter 78* covered England, Wales and Scotland in Clause 2 but did not extend to Ireland. Clause 4 provided for any local authority, or promoters having the consent of such local authority, or a road authority under the provisions of the Highways Acts, to apply for a Provisional Order from the Board of Trade, to lay down a tramway within that authority's jurisdiction. Further, when such tramway had been completed by the local authority, or it had acquired possession of any such tramway built by another promoter –

> "Such authority may, under Clause 19, with the consent of the Board of Trade, and subject to the promoters of this Act … lease, or demise to any person, persons, corporation or company, the right of user of the tramway, and of demanding and taking ….. the tolls and charges authorised".

However, and still under Clause 19 –

> 'Nothing in this Act contained shall authorise any local authority to place, or run carriages upon such tramway, and to demand and take tolls and charges in respect of the use of such carriages'

And yet again, under Clause 43 of the Act, the local authority was given the inalienable right to –

> 'Require such promoters to sell, and thereupon the promoters shall sell to them their undertaking .... upon terms of payment of the then value of the tramway …. Within six months after the expiration of a period of 21 years from the time of construction of the tramway.'

Under this inequitable and nepotistic Act, all promoters of tramways in Britain – other than Ireland – realised that in 21½ years, they faced compulsory take-over by the local authority and, of course, advancement towards improvement of methods of transport were thus totally stultified. Further, and not unnaturally, in the gathering civic socialism of the day, most local authorities (even the impecunious ones) saw it as the proof of their capability and emblem of their assertation to run their local tramways. Many had dealt similarly with the parks, water, electricity and other services and assumed enormous political power in the process. The great majority of tramways did, in fact, become absorbed into the local authority fold and thereby received their electric current from the local authority powerhouse. Those that did remain under private companies were left to negotiate for current from the local authority which charged exorbitantly. As a result, *every conceivable means of economising in current was researched and exercised.* Some went as far as going to arbitration over the excessive rates charged for electricity.

The 1870 Tramways Act functioned for over 120 years, finally only being repealed in 1993.[4]

Tramway engineers spent a great deal of thought on methods to economise in use of current. Outstanding amongst these was the knowledge of the dc tramway motor's characteristic of functioning as a generator when coasting with the power off. If this could be harnessed, not only could it be returned to the overhead power supply but it could also be utilised for electric braking at no expense at all to the tramway company. The process became known as 'Regenerative Braking'. This process was not applicable to tramways alone. Already it was being usefully applied to other electric road vehicles. These mostly comprised light personal vehicles (or motor cars) where the regenerated current was dissipated in charging the traction batteries.

The local authorities seethed under their 21½ year veto on the ability to run, and reap from the local tramways the excellent harvest of fares. The only local authority at first allowed to run its own horse tramway was that of the Corporation of

every conceivable means of economising in current was researched and exercised

Huddersfield, which, by its Act of 1882, was given this facility simply because of its hilly terrain and inability 'to demise the tramways on such terms as, in the opinion of the Board of Trade, would yield the Corporation an adequate rent'.[5]

One other tramway system had this legal right, from 1870 onwards, to run its own tramcars and take its fares, and that was the Corporation of Glasgow. This it engineered for itself in terms of the Schedule to its Act, the Glasgow Street Tramways Act, 1870, of Royal Assent 10th August 1870, Clause Third, which stated:

> 'The Corporation shall have power, within the said six months after the passing of the Act, to intimate to the Company their desire to be substituted in place of the Company ....' [6,7]

This, of course, they did, and the newly formed firm of 'Glasgow Street Tramways Company' was dissolved without ever having owned or run a single tramcar. The lines were then leased to its successor, the new company 'The Glasgow Tramway & Omnibus Company Ltd.' The hilarity behind this piece of legislation lay in that the good Corporation of Glasgow simply never realised that they had indeed engineered for themselves the privilege of running their own tramcars.[8]

Around the turn of the 19th/20th century, very much thought went into the possibility of harnessing the electric motor's potential for regeneration towards the mounting problem of the cost of local authority electricity and the possible ways of effecting economy. While many engineers were tackling the problem, as we have seen, two centres of thought predominated: that of the Johnson-Lundell Electric Traction Company of Southall and of John Smith Raworth, Electrical Engineer, at that time steeped in tramway administration.

The Johnson-Lundell firm's tramcar experiment is the subject of a separate commentary by co-author John Markham in the technical chapters. The basis of their design was not dissimilar to the later claims of Maley & Taunton of Wednesbury, near Birmingham.

THE GRAVESEND, ROSHERVILLE & NORTHFLEET SERIES SYSTEM

The series traction system first installed on the Gravesend and Northfleet line was unique in Britain being a development from a small number of experimental lines constructed in the United States. However a change from those in America concerned the way in which the motor on each tramcar at Gravesend was energised and controlled, even though a constant current supply was used in all cases.

The traction motors at Gravesend were built by Elwell-Parker and axle-hung, two-pole machines with double helical gearing to the axle. They were similar in many ways to the later Mather & Platt No.5 machines which ran for over 70 years on the Snaefell line on the Isle-of-Man.

For a dc motor to produce torque (tractive effort at the wheels), two conditions have to be fulfilled:

1. The field system has to be magnetically polarised and cause its magnetic flux to pass through the armature and its winding
2. A current must be caused to flow through the armature winding to interact with the magnetic flux

Unless both of these conditions are met no torque or tractive effort will result. The direction of the torque or tractive effort is dependent upon the relative directions of the field flux and armature current.

When stationary the 50 amps of continuous current circulated by the supply system was passed through the armature but without the field system being magnetised. No tractive effort resulted.

By means of the gradual *introduction* of a resistance into the circuit a voltage was produced across it and the field winding was connected in parallel with this resistance. Some current this produced was prompted to flow through the motor field. By increasing the amount of resistance in circuit, so the field excitation and thus magnetisation was increased. It is probable that the ohmic value of the 'shunt' resistance and the field windings may well have been of similar values, but as the field had two field coils it is possible that a series-parallel system of field coil excitation may have been adopted as a way of minimising resistance losses.

The controls available to the motorman gave him the ability to select excitation direction as well as its value. This therefore meant that he could, by applying tractive effort in reverse when under way, achieve reverse torque / tractive effort to slow the car down. This would be a regenerative brake, not by maintaining the same polarity of armature voltage and reversing the direction of the armature current, but by maintaining the armature current direction and reversing its voltage, *ie* regenerative braking.

The line had been opened for electrical running on 2nd August 1902 consisting of 2½ miles of double-track and four miles of single-track with passing loops, built to standard gauge.

THE INVENTOR — JS RAWORTH:

Raworth was born in 1846 and gravitated towards the study of electricity in its very early days of practical application. Trained at R&W Hawthorns of Newcastle-upon-Tyne as a mechanical engineer and with Wren & Hopkinson of Manchester, he joined Siemens Brothers, the famous electrical firm in the late 1870s. He worked on ships' lighting systems and power supply ashore. In 1886 he moved to London as Chief Engineer to the Anglo-American Brush Electric Lighting Corporation and as they expanded, he organised their take-over of Henry Hughes' Falcon Works at Loughborough, as Superintending Engineer. The year 1891 saw him appointed General Manager of the Brush firm just as electrification of horse tramways was getting, tentatively, under way.

John Raworth, then an electrical engineer of widespread experience, singular perceptivity, and with many registered patents, took out significant patents in 1903 for electrically propelled vehicles and their regenerative devices. These laid the foundations for his private firm to come – the Raworth Traction Patents Company Ltd, (RTP) He registered this on 1st October, 1904, following the testing of his first one-man tramcar, Southport 21, powered by only one motor.

Raworth tackled the total economy problem of the tramway industry

Raworth tackled the total economy problem of the tramway industry, especially from the aspect of the small town and rural systems most hit by expensive traction current. His approach was from three aspects: cutting down on current usage, minimising running staff and in producing small, inexpensive-to-run cars. His demi-cars, which were all fitted with regenerative braking, were the smallest tramcars in the country, all single-deckers with reversed platforms so that boarding and alighting were always at the front where the one-man driver/conductor supervised the fare collection as the passengers went past him in single file. The cars were economical in current consumption both from their small size and weight as well as from the function of their regenerative braking motors. They were ideal for lightly patronised town routes and for rural, or semi-rural lines as can be see in the plan of Southport 21.

THE INVENTION OF A MEANS OF ECONOMY — THE REGENERATIVE BRAKING CONCEPT:

Electricity was not then used in factories. All were steam or water-driven; pulley-shafts and belts distributed power in those days. As a result, power houses were largely idle by day, their sales being almost entirely for small power supplies for the likes of domestic ironing and for lighting. With the tramways being electrified, day-time consumption began to soar so that power suppliers who had a tramway customer had a most welcome improved load factor over the greater part of the 24 hours. Sadly, the generator stations, then mostly under civic or local authority ownership, did not miss the opportunity to exploit the situation right up to the hilt and jacked-up their charges very considerably. By this means the greatest current economy commanded widespread research amongst the tramway fraternity.

Raworth had put a great deal of thought to this problem of the cost of current and directed his creative mind towards harnessing regenerated current and if returned to the overhead, could be used by other tramcars of the system in preference to buying from the line at such high prices. This could exert a degree of economy for the tramway company. He invented a form of regenerative braking using in the first place a single traction motor, the compound-wound fields took the place of the otherwise universal series fields and required a much modified type of controller, looking like, and functioning in much the same manner as the well-known 'Chadburn's Ship's Telegraph'.

As chairman of the board of the Devonport Tramway Company, Raworth commenced there a series of experiments with shunt-wound motors and regenerative control. 1902 saw his first provisional registration of two patents for *Electrically propelled Vehicles and Regulating Devices* which he replaced in January and February 1903 becoming the basis of his subsequent company *Raworth Traction Patents Limited* in 1904. Following this, 9th July, 1903 saw him successfully demonstrate the first of his regenerative braking tramcars at Devonport. The car's regenerative apparatus involved the series notches, only, but a separate car, lurking within the depot, and not then quite ready for demonstration, was at that time being equipped with full series-parallel regenerating equipment.

THE DEMI-CAR —
ANOTHER MEANS OF ECONOMY:

Raworth, meanwhile, had become Technical Director on the Central Board of BET, which governed the peripheral boards of all its sub-companies. He was chairman of both Southport and Devonport as well as much else including the board membership of the North Staffs Tramway Company under the chairmanship of the well-known Emile Garcke, Managing Director of BET, and a driving force of the tramway world. Garcke was an erudite and enthusiastic person.

Raworth, meanwhile, had become Technical Director on the Central Board of BET

Raworth's prolific brain, still battling with the excessive cost of electricity, seeking answers to the onerous civic and local authority power rates, conceived a design of tramcar for the less utilised (and so less lucrative) tram routes. These were mostly suburban and country services. The standard tramcar of the day was a ruggedly built wooden-bodied vehicle of twin decks. Very many of them were by then becoming top-covered and could weigh, unladen, up to 11 tons. Fully laden, with all seats taken and a full complement of standing passengers this could increase to over 14½ tons.

SOUTHPORT TRAMWAYS COMPANY
CAR 21 OF 1903:

Raworth's design of a small single-decked saloon bodied tramcar was certainly one answer to the problem of current costs. The car was much shorter than the standard car and was of light construction. The prototype car was built by Brush in Loughborough and the Board of Trade inspection was sought by letter of 22ndJuly, 1903, following testing on the Brush experimental track. This was car 21 in the Southport Tramway fleet. It had a body length of 20ft 0in overall, featuring a 2-window saloon with clerestory roof and was mounted on a specially designed lightweight truck. It had a single Brush 800C motor, modified for regenerative braking, and was capable of 18 mph service speeds. The platforms were reversed to permit nearside front loading as previously described for one-man operation and the car weighed less than 4½ tons. All subsequent demi-cars had two motors. The semi-vestibuled platforms protected a bench seat for 3 smokers at each end while the saloons accommodated seven aside on cushioned seats.[9]

In service on the relatively level terrain of Southport it used 0.28 units of power per car-mile as opposed to 0.55 units for a double-decker. However, for an assessment of regenerative braking capability, the costs of working the car were claimed at 2.26d (approximately 1p) per mile: current costs being 0.56d, drivers' wages 1d and repairs 0.7d. The success of the Southport and Devonport cars generated a good deal of interest and comment in the technical press leading quickly to further orders. The Devonport car was not a demi-car, being a standard car of the fleet, experimentally converted to the regenerative principle before proceeding to construction of Raworth's new design.

Southport 21 was essentially an experiment much as all prototype machinery is. The single motor was not adequate even for so small a car and, as mentioned, the

John S Raworth's first experimental demi-car was Southport 21. Note the semi-vestibuling of the platform – a feature only once repeated – and the outside brake spindle. The car is seen at Birkdale Station terminus and worked to London Square on off-peak services. *(Photo: CC Hall collection)*

JS Raworth's first demi-car – Southport 21. This plan view shows seating accommodation and the designer's patent 'Bar' across the entrance to the smokers' seats on each platform. The raising of the bar acted as a 'Dead-man's Handle', stopping the car. [11]

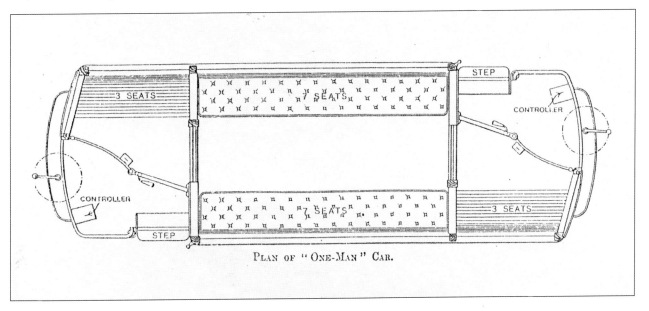

PLAN OF "ONE-MAN" CAR.

top speed on the very level track in Southport was only 18 mph. The semi-vestibuling was not appreciated by the public and was only once repeated. Apart from that, the little car was a success and more followed with an amazing variety of manipulation of minutiae until there were 23 all told, scattered around England with but one example in Scotland by the time the firm closed down on 20th November, 1911. The little Southport demi-car was probably scrapped in 1918 when the Company was taken over by the Corporation.

An outstanding feature of Raworth's design here was the emergency brake lever which worked much on the principle of what later became known as the Dead-man's Handle. This consisted of a bar, closing the passengers' entrance to the motorman's position for access and egress. The car could not be moved with this bar lifted up and so, after all entering passengers had passed through, the bar was closed down and the car was ready to move off. Should any mishap occur to the driver, a passenger could lift the bar and stop the car. This bar is clearly seen on the plan drawing of the demi-car and was an integral feature of all the Raworth demi-cars.[10]

It is little known that the inauguration of the Southport demi-car was hailed in verse of which only one stanza can now be recalled by John Markham. It went like this:

> One day from out of Birkdale
> And into Southport ran
> A tiny little tramcar
> Worked by just one man".

Of the further verses, another particularly recorded the 'Raworth Bar' safety device mentioned above, for all time!

THE GRAVESEND & NORTHFLEET ELECTRIC TRAMWAYS LIMITED – CARS 9 & 10 OF 1904:

Early in 1904 the Gravesend & Northfleet Electric Tramways Limited became the next system to order demi-cars. Two, numbers 9 and 10 of that fleet, had been built to this order by the Brush Electrical Engineering Company Limited of Loughborough, the delivery date being given as April in that year. Raworth was then a director of both the BET Group and of the Brush Company. He was also on the Board of the Devonport Tramway Company where a fellow director, JF Albright, was chairman of the Gravesend & Northfleet Company. Raworth's intimate connections within the tramway fraternity must have stood him in great stead in the promotion of sales both of his little cars and of his regenerative equipments.[12]

Raworth was then a director of both the BET Group and of the Brush Company

An improvement on the Southport demi-car, the Gravesend twins had fully vestibuled platforms and two large saloon windows. All windows were round-topped and the saloon was surmounted by a clerestory roof incorporating four opening quarter-lights. They were considerably more attractive little vehicles and were much more passenger-acceptable than the prototype, being also faster with two 27 hp Brush motors. These were rewound to shunt-wound regenerative status, working on

Gravesend & Northfleet Demi-car 9. This car, and 10, fitted with JS Raworth's regenerative braking system, were supplied to this system in April 1904. The regenerative equipment was removed in 1912 after the Rawtenstall accident.
(CC Hall collection)

series-only principle. The cars incorporated the extra comfort of full vestibuling and of running on extended (8ft 6in) wheelbase, light construction trucks which would have considerably cut down any tendency to 'jazz'. They were the only demi-cars fitted with American-type 'anti-climb' plates over the fenders to prevent persons 'free-riding' or standing on the outside fenders when in motion.

The main dimensions were:-

Saloon:	10ft 6in long x 6ft 3in wide
Length overall:	20ft 11in
Seating:	16 in the saloon plus 3 on the end platforms
Weight in working order:	5.6 tons

28% of the electrical energy used was returned into the overhead from the regeneration of the motors

Extensive tests were carried out on the hilly terrain of the system during July of 1904 and showed that on a six-mile run, 28% of the electrical energy used was returned into the overhead from the regeneration of the motors.[13]

THE GLOSSOP DEMI-CAR — NUMBER 8 OF AUGUST 1904:

Glossop 8 was delivered in August 1904 from the British Electric Car Company of Trafford Park, Manchester. It was similar to all of the Raworth demi-cars and featured a two-window saloon with clerestory roof incorporating four fan-light ventilators. The car was fully vestibuled with three drop-windows in each end vestibule. The saloon accommodated seven aside with three seats for smokers in

Below: Gravesend & Northfleet Demi-car No.10. (Photos: Walter Gratwicke, courtesy CC Hall)

each end compartment. The car had the usual Raworth regenerative equipment comprising special controllers and two Brush/Raworth shunt-wound motors. It took over operation of the Whitfield route during August, the conductor being dispensed with and the motorman collecting the fares.

The Glossop line was one of the smallest in the country. Only Kidderminster, Wrexham and Ilkeston had weekly receipts in the same category as Glossop's, varying from £100 to £135 with an average of about 8d (just over 3p) per car-mile. The Glossop system served a population of roughly 22,000 people and was run by the Urban Electric Supply Company under the management of CE Knowles. [15]

It is of interest that Glossop 8 was thought to have been the only Raworth demi-car to have been entirely built by the British Electric Car Company. The firm had been under voluntary liquidation since 7th December, 1903. Its policy of undercutting competitors' prices had back-fired, yet from reports in the Electrical Review it appeared to have been still manufacturing in June 1904 and probably for some time after that.[16]

Glossop 8 had been completed at the Trafford Park Works and another three – Barrow numbers 13 & 14 and the Yorkshire Woollen District Company's 59 – were probably laid down there only to be overtaken by the firm's closure proceedings. Raworth managed to get his old firm of Brush, of Loughborough, to agree to complete them and he apparently had to infiltrate the closed firm's shops to salvage and release the bodies already constructed (and parts) for their completion at Loughborough. This explains the good deal of indecision over exactly where they were built. They were the last of Raworth's series-only motored demi-cars. [17]

From the illustration it can be seen that the car had a very neat, although perhaps a trifle square-cut design. Apart from a much later example it was the last of the large, two-window saloon type and featured a BEC-constructed lightweight truck described by Ian Yearsley as:

'Not so much a truck, as a set of railway wagon 'w'-irons, with a simple framework of bars to support the motors.' [18]

Glossop 8, as built: one of the early JS Raworth demi-cars and the only one completed by BEC. It was fitted with Raworth regenerative equipment and was built in 1904. The gentleman with the bowler hat is Mr Knowles, Manager, and there is reason to believe that his accomplice may be Raworth's son.
(Photo: Walter Gratwicke, courtesy CC Hall)

Certainly, an excellent photograph of the Glossop car taken at a later period does show the replacement of this truck with a heavier, pressed-steel frame variety, perhaps suggesting that the original truck had been of too light a construction for the exigencies of service.[19]

The main dimensions of the Glossop car were:-

Length, overall	20ft 11in
Length, over platforms	19ft 8in
Platform length	4ft 7in
Body (saloon) length	10ft 6in
Width, overall	6ft 9in
Height (rail to sills)	2ft 2½in
Inside height	7ft 6in
Total height (rail to trolley plank)	10ft 10½in
Wheelbase	6ft 0in
Diameter of wheels	31¾in
Seating: Longitudinal 16 + 3 smokers on each platform	
Total = 22	

The front cover of *Glossop Tramways* by BM Marsden has a superb picture of 8 at Whitfield terminus some time after it was new. The truck shown accommodates slipper brakes and is similar to the truck shown on Drawing 162 for the Yorkshire Woollen District. What is not visible is any form of controller though the conventional cranked handbrake is quite clear. This latter, too, must have been a modification. The photograph of the car as delivered shows a horizontal brake wheel for the motorman, again as per the Yorkshire Woollen District drawing. There are no track brakes and apparently only one traction motor. The controller looks like the 'ship's telegraph' (face plate) variety. The posed motorman appears to have his left hand on it. The man on the steps was Mr Knowles, the Glossop Manager. It is suspected that the changes to car 8 took place fairly early in its life as the Whitfield route had a continuously rising gradient for about half its length.

Marsden's text includes:

'(the car).....utilised Raworth Regenerative equipment comprising controllers...and Brush-Raworth shunt-wound motors'.

He also says that a BEC truck was supplied but the delivery photograph shows a truck unlike the usual 'SB' range of trucks made by BEC. Their trucks were normally made from pressed steel sections rather than the bar form of construction shown in the picture. A replacement truck with two motors at an early stage seems likely. It is suspected that the motors were both shunt-wound, as Marsden says, but were connected permanently in series (*ie* no transition), the fore and aft Raworth controllers being retained. [20]

Here BM Marsden states 'There are no track brakes and apparently only one traction motor'. This is debateable – Ian Yearsley giving 'The first demi-car had only one motor, a modified Brush 800c, although all later demi-cars had two'.

HALIFAX CORPORATION – CARS 95 & 96 – EARLY AUTUMN 1904:

Halifax Corporation's tramways had to contend with one of the hilliest terrains in England and ordered two demi-cars for its shuttle service along Horton Street to the Old Station which sloped steeply down to the station tram terminus. Delivered in the autumn of 1904 they were first used on the fairly level Skircoat route to train motormen in the handling of regenerative braking cars. It is reported that in 1906 they were involved in a run-away collision, the result of a powerhouse failure in the overhead current neutralising the regenerative braking by cutting off the regenerative circuit.

Laid-down and largely completed by BEC, they were 'rescued' from the failing (and folding) company by the Brush Company and completed at Loughborough. There is, in fact, sufficient incidental evidence to be certain about the bodywork construction of these cars, as Ian Yearsley has stated that:-

'It has been suggested that these cars were originally cross-bench in construction but evidence points to the contrary. After the 1906 collision they were repaired and a 1907 photograph shows one of them in the standard demi-car condition. On 11th August, 1910, the Halifax Town Clerk wrote to the Board of Trade informing them that the two cars had been converted to cross-bench form, and enclosing an official photograph of 96, requesting BOT approval. One of them was involved in an accident in 1910 when an elderly man alighted from the off-side of the car and was struck by a car on the opposite track'.[21]

An excellent official photograph – obviously the one sent to the Board of Trade, shows Car 96 as a totally open-sided, crossbench-seated vehicle with running board access. The cut-away saloon side showed the remnants of the original two-window saloon construction in the form of two round topped window frames with the common central pillar removed indicating that originally they were built as two-window saloon cars. This is exactly what would have been expected from the Raworth design in its initial phase of construction.[22] The mere fact of an open-sided car with a clerestory, four quarter-light opening ventilator roof indicated reconstruction from standard saloon configuration!

Another interesting photograph from the late Bob Parr collection shows demi-car 95 sitting on trestles outside some depot with a standard Halifax Corporation Peckham Cantilever truck in the foreground. The juxtaposition of both could suggest the marrying of the two but the outstanding feature is that 95 has been rebuilt with a three-window saloon, with flat topped quarter-glass ventilator panes, conforming to later Raworth designs. Sadly, the date of this photograph, shown here, is not known.

Car 95 was rebuilt yet again in 1918 as a mobile kitchen and for this requirement, roof-trusses were installed, on the cantilever principle, probably with central turnbuckles for adjustment of tension. Number 96 was used during the First World

Halifax 96 as an open-sided car.
(Photo: Courtesy HN McAulay)

War as a mobile Army Recruiting Office. Both cars were later converted to works cars and finally scrapped in 1926.[23]

The scant general dimensions available were:-

Saloon length:	10ft 6in
Saloon Breadth:	6ft 3in
Gauge:	3ft 6in
Seating Capacity:	16 in saloon + 3 smokers' seats on each platform
Cost:	£500

The Halifax truck was an ordinary Peckham Cantilever fabricated truck, fortified by the super-addition of twin strengthening bar extension to the main side frame sandwiching the crowns of the axle-box horn-blocks. The additional low-slung weight would have helped to stabilise sway on this 3ft 6in gauge system. Halifax was a hilly town and known to have upwards of eight blow-overs in high winds! One of Halifax's two demi-cars, 95, is seen here rebuilt with three windows in the saloon, standing on trestles awaiting a truck – query the one in the foreground?

Halifax 95 as a mobile kitchen, offering rice pudding for 1½d (0.42p) (Photo: BJ Cross collection

THE FIRM OF RAWORTH'S TRACTION PATENTS – 1ST OCTOBER, 1904:

The firm of Raworth's Traction Patents was launched on 1st October, 1904 with a board of management involving colleagues from the BET and several of its sub-companies, such as the Birmingham & Midland, Devonport, Dudley – Stourbridge and Kidderminster. Also included were one or two directors and engineers from interested general electrical firms. Its be-all and end-all centred upon the manufacturing of regenerative braking equipments for sale to equip established standard cars as well as the design and equipment of the now-becoming-famous little demi-cars. These were all within the parameters of the conservation of electrical current leading towards the economical running of tramway systems requiring to purchase power from local authority sources.

At least two conventional Devonport cars had been fitted up with Raworth's regenerative equipment for experimental and development purposes and some half dozen demi-cars had already been delivered before the floating of Raworth's Traction Patents. These comprised the Southport car, two each to Gravesend & Northfleet and Halifax Corporation, while the new firm's prospectus mentioned the Glossop demi-car as already having been delivered.

'The Demi-car is designed to save half the labour and half the energy required by a large car;'

Raworth's publicity lives on in the Exhibition Hall of the National Tramway Museum at Crich.

(Photo: HN McAulay)

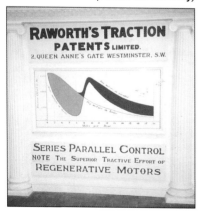

The Tramway and Railway World.

[DECEMBER 17, 1904.

Notice of Prospectus now being issued dated 8th December, 1904, and duly filed with the Registrar of Joint Stock Companies.

RAWORTH TRACTION PATENTS (Limited).

CAPITAL ... ... £30,000.

Divided into 15,000 Preferred Ordinary Shares of £1 and 15,000 Deferred Ordinary Shares of £1.

15,000 PREFERRED ORDINARY SHARES ARE NOW OFFERED FOR SUBSCRIPTION.

They are entitled to a fixed cumulative dividend of 6 per cent. per annum, and to participate *pari passu* with the Deferred Ordinary Shares in the surplus profits of each year remaining after paying the fixed dividend for such year on the Preferred Ordinary Shares, and a like dividend for such year on the Deferred Ordinary Shares.

The Preferred Ordinary Shares and the Deferred Ordinary Shares rank *pari passu* as regards capital.

The Directors have power to raise on Debentures or Debenture Stock any amount not exceeding the nominal Capital of the Company.

Payable as follows :—2s. 6d. per share on application—7s. 6d. on allotment.

The Balance, as and when required, in calls not exceeding 2s. 6d. per Share at intervals of not less than two months.

DIRECTORS :

C. SHIRREFF B. HILTON, of 41, Roland Gardens, Kensington.
(Chairman of the Birmingham and Midland Tramways, Ltd.)
THOMAS BROWETT, of 9, Oakwood Court, Kensington.
(Late Joint Managing Director of Browett, Lindley & Co., Ltd., Engineers).
G. F. METZGER, of 3, York Street, Manchester.
(Late City Electrical Engineer, Manchester).
A. K. BAYLOR, of Trafford Park, Manchester.
(Managing Director British Electric Car Co., Ltd., Manchester).
J. S. RAWORTH, of 2, Queen Anne's Gate, Westminster.
(M.Inst.C.E., M.I.E.E., M.I.M.E., Director of the British Electric Traction Co., Ltd.).

BANKERS :

LONDON & COUNTY BANKING CO., LTD., Lombard Street, London, E.C.

SOLICITORS :

BRABY & MACDONALD, 5, Arundel Street, Strand, London, W.C.

SECRETARY AND OFFICES (pro tem.) :

JOHN I. HALL, 2, Queen Anne's Gate, Westminster, S.W.

ABRIDGED PROSPECTUS.

This Company has been formed to acquire the British Patents granted to the Vendor, Mr. J. S. RAWORTH, of 2, Queen Anne's Gate, Westminster, M.Inst.C.E., for the system of automatic regenerative control of electrically propelled vehicles and for the Demi or one-man car.

The general principles of the inventions are

1. To provide a motor for use upon an electrically-driven tramway car or railway train which utilises the impetus of the car, either when descending a gradient or when reducing speed to regenerate electrical energy which passes back to the trolley wire.

2. To provide an electrical tramway car, which, while complying with Board of Trade requirements, is suitable for working by one man only.

The price to be paid for the patents is £18,600, payable as to £3,600 in cash and as to £15,000 by the allotment of 15,000 Deferred Ordinary Shares credited as fully paid up, and includes the transfer of all uncompleted orders as from the 1st October, 1904, patterns and drawings. Of the purchase money £100 is payable for the goodwill.

The invention is applicable also to electric railway trains, and to electric automobiles.

The Demi-car is designed to save half the labour and half the energy required by a large car; it is principally applicable to the numerous lines on which the traffic is light and the profits are small.

THE FOLLOWING CARS EQUIPPED UNDER THESE PATENTS ARE IN SERVICE.

Southport Tramways Co.	1 demi-car with regenerative control.
Devonport Tramways Co.	6 large cars, regenerative control.
Halifax Corporation	2 demi-cars with regenerative control.
Gravesend Tramways Co.	2 demi-cars with regenerative control.
Glossop Tramways Co.	1 demi-car with regenerative control.
Scarborough Tramways Co.	3 large cars, regenerative control.
	15

THE PATENT EQUIPMENTS FOR THE FOLLOWING CARS ARE ON ORDER.

Southport Tramways Co.	1 large car conversion to regenerative control (Second Order).
Barrow Tramways	2 demi-cars with regenerative control.
Bournemouth Corporation	1 large car conversion to regenerative control.
Yorkshire Woollen District Tramways Co.	8 large car conversions to regenerative control.
Devonport Tramways Co.	1 demi-car with regenerative control.
	7 large car conversions to regenerative control (Second Order).
Gravesend Tramways Co.	5 large car conversions to regenerative control (Second Order).
Birmingham & Midland Tramways Co. ...	40 large cars, regenerative control.
	65

Negotiations for several other orders from corporations and companies are in an advanced stage.

Letters Patent for the system of regenerative control have been granted to J. S. Raworth in the United States of America and the Certificate of Allowance has been issued in Germany. In both these countries official searches for anticipations are made.

Mr. J. S. RAWORTH has consented to act as Managing Director of the Company for the present.

The above extracts from the prospectus are published for information only, and not as an invitation to subscribe for shares. Full copies of the prospectus upon which alone application will be entertained, can be obtained from the Bankers or Solicitors or from the Secretary at the registered address of the Company.

Reference was made to three others already on order: two for Barrow and one for Yorkshire Woollen District. Over and above this, complete regenerative equipment had already been delivered for the upgrading of existing standard cars, including:-

Devonport	6	Bournemouth	1
Scarborough	3	Yorkshire Woollen District	8
and on order for:		Devonport	7
		Gravesend	5
Southport	1	Birmingham & Midland	40

So, the new firm was getting off on an excellent footing with all those orders through the 'Old Boy' Network'. [24]

John S Raworth's proposed Company Prospectus was issued late in 1904 in the name of Raworth's Traction Patents Limited with a capital of £30,000. The Company was formed on 1st October, 1904. Half the capital was in preferred ordinary and half in deferred ordinary shares. The Prospectus also mentioned Glossop 8, already built, trucked and delivered the previous August by the British Electric Car Company, and in liquidation at that time. It stated that his son, Alfred Raworth, had joined his father's firm in 1903 having completed his apprenticeship elsewhere. Alfred's job, initially, was to deliver new demi-cars and teach the motormen how to handle them.

His subsequent career, after the Company had been wound up in March 1912, was with the London & South Western Railway Company as Chief Electrical Engineer, subsequently taking the same position on the South Eastern & Chatham Railway, leading to his becoming Chief Electrical Engineer with the amalgamated Southern Railway in 1923.

THE REGENERATION IN HECKMONDWIKE:

The Yorkshire (Woollen District) Tramway Company, a BET concern, operated tramways in the Dewsbury and Batley areas with a number of routes serving the neighbourhood. It did not have its own power station but purchased its traction power from various authorities for different sections of line.

In particular its route to Hightown was supplied with power by the Heckmonwike Urban District Council. The consulting engineer to the council had advised that a supply of 100 kW would suffice for this section of tramway and the traction feeder circuit-breaker at the power station was set to this limit. In practice, if one assumes a line voltage of 500 Volts, this would be around 200 Amps.

The route in question had a number of gradients and the company found that it could only operate two cars on it, one going up and one going down, without the 200 Amp current limit being exceeded on a regular basis. Heckmonwike steadfastly refused to increase the maximum power available so the tramway company had to investigate other means of addressing the problem.

In 1904 it took delivery of regenerative demi-car 59 and proved for itself that its power consumption per car mile was much less than the single-deck four-wheel cars that were running on the service, and forthwith ordered 8 sets of Raworth regenerative equipment, six for single-deckers and two for double-deckers. The demi-car was transferred to Barnsley.

The single-deckers were put to work on the Hightown route and were successful in enabling more cars to operate in section without the regular power failures that previously occurred.

Heckmondwike complained somewhat bitterly that their power station meters were then giving erratic readings as a result of the regenerative cars

Heckmondwike complained somewhat bitterly that their power station meters were then giving erratic readings as a result of the regenerative cars, which the tramway company seems to have ignored. Dewsbury's reaction was rather different; they demanded an assurance from the tramway company that it would indemnify the Corporation from any matter arising from their use of regenerative tramcars.

It was 1909 before Heckmonwike was able to meet the tramways need for additional power, at which time the Woollen District Tramways converted their regenerative cars to normal series motor operation.

The Yorkshire Woollen District Tramway Co., Ltd., –
Demi-car 59 (I) of February 1905:

W Pickles gives this demi-car as built in 1904. [25] Ian Yearsley indicates that the car was built in February 1905. [26] Both may be correct if the two dates should represent the departure from the Trafford Park Works and the completion date at Brush Electrical Engineering in Loughborough. By this time Raworth had been working on a new type of controller that would introduce regenerative braking to the traditional series-parallel motor. This was the 'R6' controller, the most distinctive feature of which had been a 'lost-motion ring'. This allowed one of the contacts to lag behind when the controller handle was reversed and gave a different set of connections when notching down from those given in notching up. [27,28]

The dimensions of 59 were:

Length:	20ft 6in
Overall width:	6ft 6in
Seating:	14 passengers in the saloon + 3 each on the platforms for smokers, making a total of 20.

There were two Brush, shunt-wound, series moors of 17hp and the truck is said to have been of BEC-build with a wheelbase of 6ft 0in. There remains this dubiety about the date of building. Pickles has stated that on account of the economy in power consumption by 59's regenerative braking, Yorkshire Woollen District placed a further order on 28th September, 1904 for eight sets of Raworth regenerative equipment for fitting to existing standard cars in their fleet. This would require 59 to have been built at least in the first half of that year. [29]

The demi-car did not operate for long in the Yorkshire Woollen District Company's fleet. Having evidently done its job of increasing the patronage on the Heckmondwike to Hightown route to the point where larger cars became necessary, it was transferred to the Barnsley & District Electric Traction Company for £500, under the same management. It became 13 in that fleet and was seen in Barnsley Depot in

1926 by Charles C Hall, of Sheffield. Yearsley states the Yorkshire Woollen District car, or its consorts, the Barrow-in-Furness demi-cars, were the first of the genre to be built with three-window saloons. Apart from six Plymouth (Devonport) vehicles that reverted to the two-window saloon pattern later, the three-window design was continued until the end of the series.[30]

THE BARROW-IN-FURNESS DEMI-CARS 13 & 14 OF 1905:

Laid down in the works of the BEC Company, and probably completed in 1905 by the Brush Company of Loughborough, it would seem likely that the majority of the work had been done at Trafford Park. Ian Yearsley gives the *Tramway & Railway World* directory section for July 1905 as describing Barrow's fleet as being by Brush and BEC, and mentions two demi-cars. The system gauge was 4ft 0in, having been the gauge of the previous steam trams.

The Barrow system was operated originally by the BET which would later transfer it to the Barrow-in-Furness Corporation on 1st January, 1920. The two series-only regenerative braking demi-cars with Brush 800 (?) 17 hp motors were put to work immediately upon delivery on the Ramsden Dock – Roose route during certain hours when the traffic was light. The Company's name was listed among others on the Raworth's Traction Patents Ltd stand at the 1905 Tramway & Railway Exhibition in London.

Facing page: Yorkshire Woollen District Electric Tramways Ltd. 59 (I) was bought for the Cleckheaton – Moorend route. It is seen here at Milbridge. Having been bought in December 1904, it was sold in the following year to the Barnsley District Electrical Traction Company, becoming their 13 for the little-used Smithies Route. (Photo: CC Hall collection.)

John Smith Raworth's General Arrangement Drawing of the Yorkshire (Woollen District) demi-car 59 (I) was one of the second group of demi-cars. Building was commenced by BEC and completed by Brush. This shows the standard three-window saloon body but with the last of the Brush shunt-wound motors, in a lightweight BEC truck. The drawing shows Raworth's 'dead-man' type bar on the platform, separating off the motorman's stance from the three seats for smokers. (Illustration: Ex-Yorkshire (Woollen District) File, courtesy CC Hall collection.)

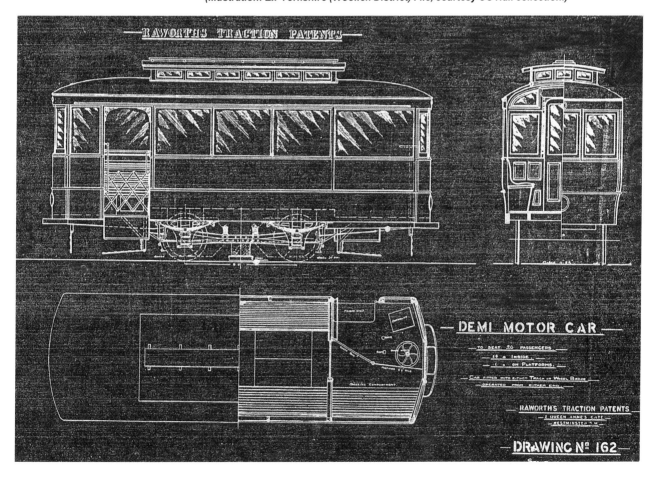

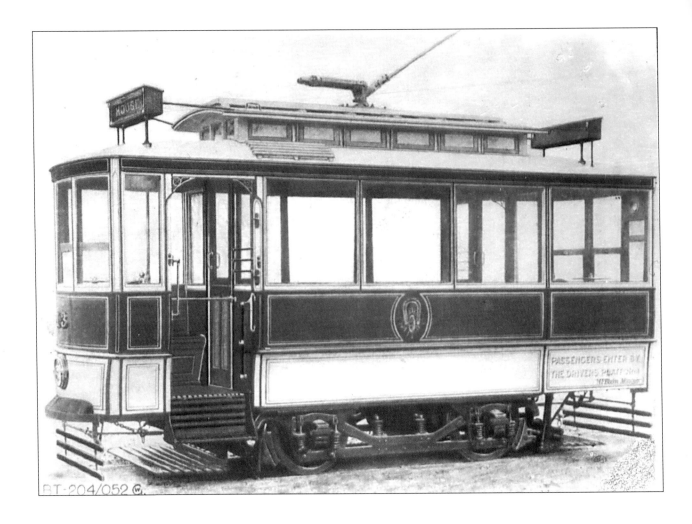

Barrow-in-Furness demi-car 13 is illustrated in this BEC Company official photograph taken in their Trafford Park Works yard. Headboards were later fitted for advertisements as can be seen in the following view. *(Photo: Walter Gratwicke, courtesy, Ian L Cormack)*

Barrow's 22-seater demi-cars took 0.65 units per mile as against the 1.1 units consumed by the ordinary cars. Their cost per mile was 3d compared with 4.78d per car-mile of the standard cars. Thus, by using the demi-cars there had been a profit of 1.12d per car-mile compared with the 0.66d loss per car-mile incurred by the running of standard bogie cars. The demi-cars ran for nine hours each day of service and were scrapped around 1919-1920, beyond repair, when the Corporation took over from BET. [31]

From photographs, these cars were of the three-window saloon type with six opening glass fan-light ventilators set into a clerestory roof. They also had all enclosed vestibuled platforms with drop-light windows. Their very light construction trucks attributed in design to BEC are best seen in the photograph of the Glossop demi-car and they were the last of the series-only type. [32]

CHESTER CORPORATION TRAMWAYS DEMI-CAR 13 OF 1905:

Built by the Brush Electrical Engineering Co., Ltd., Loughborough, in 1905, this was the first of the compound-wound demi-cars with Brush compound motors. It was also the first fitted with the Brush/Raworth new series-parallel Type R6 regenerative braking controllers with lost-motion ring. Number 13 was a slightly bigger 20-seater car with three square window saloon and a clerestory roof incorporating six quarterlight ventilators and cost £520. The Chester gauge was 3ft 6in, probably on account of the very narrow streets. The car was apparently not very much used as passenger demand required larger cars. It was later used as a works car but was believed to have existed right up until the time of closure of the system on 15th

Here is Barrow-in-Furness 13 again, on the Roose route at Ramsden Square circa 1916. *(Photo: courtesy Ian L Cormack)*

February, 1930. The body, however, was in existence until 1948 when it was destroyed.

While following the now established basic Raworth design, yet so many detail differences did appear that it was probable that the Brush firm agreed to build the car providing they could use standard parts common to other single-deck cars. The result was a 22ft 6in overall length – two feet longer than the Yorkshire Woollen District car, and it was provided with a more conventional truck design. The photograph shows a totally enclosed three-window saloon and platforms enclosed by three drop-light windows.

Chester 13 had been bought to function as a workman's car. It was then used as a stores van but was pressed into passenger service on race days. It had originally seen service on the Torvin Road and the Christleton Road routes running from Broughton Fountain. [33]

ERITH URBAN DISTRICT COUNCIL DEMI-CARS 15 &16 OF NOVEMBER 1906:

By 1906 Erith had been losing heavily on its Northend shuttle service and determined to try demi-cars in an attempt to diminish this loss. Two Milnes, Voss & Co., Ltd. bodies with special Type 40 Mountain & Gibson trucks fitted with Raworth Traction Patents Ltd regenerative braking Westinghouse motors were delivered in November, 1906, at a total cost of £1,346, the pair.

They were placed in service on a 10-minute shuttle service in January 1907, running in the afternoons only, after 1.00 pm. They made no profit, probably naturally because passengers had to change cars at 'The Wheatley'. This was rectified in the following year with the demi-cars running through on the main line to Belvedere Station but failed and was stopped within three months. Obviously the Northend service was not viable and after a certain amount of playing about with it, 16 was hired out to the Dartford Light Railways late in 1915 and purchased outright in 1916, only to be lost in the Dartford Depot fire of 8th August, 1917. Number 15 was sold to Doncaster Corporation Tramways in 1917, becoming their 37.[34] There is some dubiety as to these disposals as Yearsley gives Erith 16 as being sold to Doncaster.[35]

The dimensions of these two demi-cars were:-

Length (overall):   22ft 6in
Length (over dashes):   21ft 6inLength (over pillars):
Length (over pillars)   11'-0'
Width (overall):   6ft 8in
Height (rail to trolley plank) 10ft 5½in
Trucks:Mountain & Gibson (special) Type 40, 5ft 6in wheelbase
Seating: 20 persons [36]

Next to nothing is known about Doncaster 37 other than that it ran on the Avenue Road route for some years before withdrawal.[37]

THE DARWEN CORPORATION TRAMWAYS DEMI-CARS 14-16 OF 1905 & 1906:

Darwen, south of Blackburn, ordered a single demi-car for its steeply graded Hoddlesden route in 1905. Their 14 was delivered in November of that year and was sufficiently successful for a further order of two more,15 and 16 to be placed and delivered in late 1906.[38] (Another source gives, alternatively, **15's** delivery as November 1905 followed by 16 and 17 in late 1906.[39])

Whichever way their numbers went, the first one was earmarked for the Hoddlesden route while the other two were allocated to the Bolton Road route. Little has been recorded of this out-of-the-way system other than that they were all retired from regular service in 1912, probably as a result of the Board of Trade's opinion of Raworth's regenerative brake system following the 1911 Rawtenstall accident. They may have worked odd turns, when required, but by 1919 they were no longer included in returns made to Garcke's Manual. The original one became Works Car 1 in 1912 and all were offered for sale in 1922. One car body was sold in 1926, another became a shelter at Sett End on the Hoddlesden route (which closed in 1936) and the third car was broken up.[40]

They, and the Erith couple, were longer than the Raworth standard demi-car being 22ft 0in in length as opposed to the previous 20ft 0in which allowed for the fitting of more standard trucks.

Builder: G.C.Milnes, Voss & Co., Ltd.
Type of Car: Demi-car: 22-seater with 3 rectangular saloon
               windows and 6-glass ventilator clerestory
Truck: Mountain & Gibson Type 21EM
Motors: 2 x Westinghouse (? hp) modified for Raworth
               regenerative braking equipment.

PLYMOUTH CORPORATION TRAMWAYS DEMI-CARS 37-42 OF 1906:

These were built by Brush in 1906 and – shades of Southport's 21 of 1903 – featured half-open vestibules: a retrograde step! They were neat enough little cars, of two rectangular saloon window construction, each having two fan-light panes and ventilation scoops set into the cant rail – more typical of the Preston design of car body. The maker's photograph shows what appears to be the thin edge of a clerestory roof but it is much more likely that this was the edge of the trolley plank.[41]

There is little that has been written of these cars but Yearsley rated them as a

"fair degree of success, at Plymouth, working originally from Pennycomequick to the Hoe, and later from Friary to Drake Cross".

Builder: The Brush Electrical Engineering Company Ltd.,
               Loughborough, 1906
Type: 20-seater demi-car with 2 rectangular window saloon,
               half-open vestibules

34

Truck: Brush Type AA
Motors: Westinghouse re-wound for regenerative braking [42]
Controllers: Raworth

Car 42 was converted into a welding car in 1922 surviving until the autumn of 1946 – more than a year after the closing of the system. [43]

They had been bought originally for the North Road Station route with its bizarre terminal arrangements but they were withdrawn from this in 1915. Numbers 37-41 were withdrawn in 1924. [44]

DIFFICULT TIMES FOR RAWORTH:

Johnson-Lundell could not prove that Raworth had usurped their inventions

On 16th August, 1907, the Johnson-Lundell firm of electrical engineers saw fit to take legal action against the Raworth's Traction Patents Limited Company for alleged infringement of patents. This is fully dealt with in the technical chapters and in this chapter's overview, but suffice to say here that it is difficult to see grounds for this as the two firm's machinery were quite different and Johnson-Lundell could not prove that Raworth had usurped their inventions. Nevertheless the lawyers went to town on it with proceedings dragging on until October 1908 by which time Johnson-Lundell had to mortgage their works to pay for legal expenses and withdrew their action.

The Raworth Company, also feeling the pinch of legal costs and loss of trade and goodwill, reduced its staff and moved from the original office in Westminster to less pretentious premises in 22 Cooper Street, Manchester.

A major order for equipment came in 1908 from Rawtenstall which was in the throes of converting to an electrified system. They required regenerative equipment for the entire fleet of 16 new double-decked cars. Despite this and two demi-car orders comprising Greenock 40 (the only Scottish demi-car) and Maidstone 18, in 1909 – the very last of the demi-cars – all this business was inadequate to keep the Raworth organisation afloat. Stringent restrictions in September 1909 were followed by the dreadful Rawtenstall collision of 11th November, 1911, and led to the Raworth's Traction Patents Limited company's resolution of 15th December, 1911 to wind itself up, voluntarily, with the final meeting being held on 14th January, 1913. [45]

Rawtenstall allegedly had another tramway accident early in 1910 on the Crawshawbooth route, involving a single car that overturned and may have decided the Rawtenstall Corporation to install slipper brakes on their cars and could have avoided the accident of 1911, but wait!

John Markham recently visited the Rawtenstall Library to research local records on this accident and had the good fortune to meet a certain Peter Deegan, of Rawtenstall, who was researching the local Council Minutes and had completed from 1891 to 1923. As a local man, Mr Deegan had no recollection of any tramway accident on the Crawshawbooth route. He immediately checked through the 1910 Minutes as well as the 1910 local paper *The Free Press* and there was no reference at all of any accident during that period. Markham could only conclude that there never was an accident to a tramcar on the Crawshawbooth line 'in the early part of 1910' as RW Rush states in *Tramways of Accrington 1886-1932*. [46]

JS Raworth was given a seat on the North Staffs. Board in October 1915 but died on 24th March, 1917. Obviously the Johnson-Lundell legal proceedings followed by the Rawtenstall disaster and then the winding up of his Company had told heavily on Mr Raworth and he died in his 75th year.

THE GREENOCK & PORT GLASGOW TRAMWAYS CO., LTD.
DEMI-CAR 40 OF 1908:

This penultimate demi-car was built in 1908 by the United Electric Car Company Limited of Preston, having changed its name in 1905 from the Electrical Railway & Tramway Carriage Works Limited. Car 40 was the only demi-car to work in Scotland and in April 1909 it was running a half-hourly service from Gourock Pierhead to Ashton. 1912 saw the car heavily decorated to mark the centenary of Henry Bell's P.S.'Comet', the first regular passenger carrying steamship in Europe which was built in Port Glasgow. [47]

Particulars:

Builders:	The United Electric Car Company Ltd., Preston
Type:	Demi-car, 3 square window saloon with monitor roof incorporating six tilting quarter-lights; full-height trolley mast.
Truck:	Brill type 21E with coiled instead of semi-elliptical end springs.
Gauge:	4ft-7¾in
Motors:	Westinghouse (? hp)
Length (body):	11ft-0½in
Length (over platforms)	22ft-6in
Width over pillars:	6ft-4½in
Clear height inside, centre:	7ft-6in
Height to trolley plank:	10ft-7½in
Seating capacity:	14 in compartment + 6 in vestibules
Interior finish:	Oak and ash

There is a thin layer of dubiety regarding dates of transfer of the Greenock & Port Glasgow demi-car 40 to the Rothesay Tramways, for it is known that reciprocation of both tramway materials and of tramcars occurred from time to time between the two closely neighbouring BET firms on either side of the Firth of Clyde. Indeed both were under the Management of Archibald Robertson, as was the Airdrie & Coatbridge Tramways – the third BET Scottish Company. The County Librarian at Rothesay for the County of Bute, Arran and the Cumbraes gave 'There were changes during the 1914-1918 War .... One small Greenock car came to Rothesay,

Greenock & Port Glasgow Tramways Company Limited 40, of 1908: official United Electric Car Company view. This was the only Scottish example of a JS Raworth demi-car.
(Photo: Author's collection)

This is the only known view of Greenock & Port Glasgow's 40 in service, running down from Gourock to Ashton.
(Photo: Struan Robertson collection, courtesy the late Ronald B McKim)

Rothesay Tramways Company 22 – the only single-truck toastrack car in the fleet, having been cut down from the demi-car No.21 ex-Greenock & Port Glasgow 40. This was the only example of a JS Raworth regenerative braking demi-car in Scotland. By the time of this reconstruction the regenerative braking feature was unlikely to have been retained.

(Photo: Author's collection)

at the same time, or a little later.' [48] He refers to the Greenock demi-car, 40, which, with a change of gauge from 4ft-7¾in to 3ft-6in, became Rothesay 21. He then proceeds to countermand himself later by stating it was 'Transferred to Rothesay from Greenock after the First World War'. [49]

Ian L Cormack, in a more formal history of the Rothesay Tramways Company, gives 1916 as the year of the transfer, with 21 becoming the only Rothesay four-wheeled tramcar. How standard gauge motors were accommodated within the narrow confines of a 3ft-6in gauge truck is questionable as is the likelihood of regenerative braking by the time the transfer was effected.

In March 1920 an estimate of £450 was made to build 'An open body for the small car (the demi-car) which came over from Greenock in 1916'. This rebuilt demi-car was observed in service in 1923. [50] Elsewhere, Cormack gives the reconstruction date as 1920 and bearing the new Rothesay fleet number of '22', [51] while Yearsley quotes 1924. [52]

The snapshot of 40 in Greenock came to the author some fifty years ago, courtesy of the late Ronald B McKim and the car is seen cruising down from Gourock towards Ashton on a dull, rainy day. Why the rear door is open is not known as it ought to be shut when travelling in that direction. The appearance of the car should be compared with the view of 22 in Rothesay.

Meanwhile, what of the body of demi-car 21? The intact bodywork of the demi-car most probably remained in store at Rothesay's Pointhouse Depot and Works, naturally retaining its number '21', for who would be bothered to paint it out? As such it would have remained until the decision was taken to transfer it to Ettrick Bay terminus as a combined residence for a staff member during the summer months, and as a store shed, possibly during the off-seasons. It long outlasted the Rothesay Tramways which closed on 30th September, 1936. [53]

The balance of choice amongst the possible dates of the demi-car arriving in Rothesay would seem heavily weighted upon Ian Cormack's information, on the two points of use of primary source material and of recorded observation. The anomalous application of the number '22' to the rebuilt demi-car while the fleet number lay unused, has long lain misunderstood but the above proposal would seem to account for it rationally.

MAIDSTONE CORPORATION TRAMWAYS 18 OF 1909:

Maidstone, having found its Tovil route seriously under-exploited, ordered a demi-car to replace economically the heavier standard cars. Delivered in 1909 and put to work on the Cannon – Tovil route, it was to be the last demi-car built.

Delivered in 1909 . . . it was to be the last demi-car built.

Rothesay demi-car body 21 in use as a store-house and temporary living quarters for outdoor staff at Ettrick Bay terminus during the summer season. This was in 1943; seven years after the tramway system had been eliminated. *(Photo: STTS collection)*

<pre>
Builders: The United Electric Car Company Limited, Preston, 1909
Type: Demi-car – 3 square, opening saloon windows, clerestory
             roof with six opening quarter-lights
Truck: Mountain & Gibson 21EM, 5ft 6in wheelbase, 31in wheels
Motors: 2 x Westinghouse (? hp) Raworth compound-wound
Controllers: 2 x Westinghouse Type T1R (Raworth) controllers
Seating: 16 in saloon, on 2+1 transverse seating, plus 2 on each
             platform = 19
Dimensions: Length over platforms: 21ft 6in
             Length over saloon: 11ft-0½in
             Breadth of body: 6ft 2in
</pre>

Maidstone was obviously satisfied enough with the Raworth regenerative braking control on the demi-car to order three more sets of the equipment for fitting to their standard cars. In 1919, as post-war traffic was building up, 18 was withdrawn to storage at the back of the depot, and never used again, being withdrawn in 1928, among the first to go. It disappeared into oblivion on the back of a low-loader lorry, and was forgotten about until 1970 when it was discovered at Winchelsea in use as a holiday caravan, more than somewhat overgrown! It was removed to a safe place in London in 1971 for storage pending restoration. Ian Yearsley was in the team which discovered and reclaimed this car which is now in the Dover Transport Museum, Old Park, Whitfield, Dover. [54,55,56] It now has a truck from a Lisbon tramcar but the body is in a dire state. Restoration is proposed but nothing has been done for several years. When formal enquiry was made by David J Holt concerning progress it was stated "We'll be doing something soon".

An interesting feature about all one-man operated tramcars highlighted by John Markham was to the effect that he had never seen in print the arrangement of the hand-brake linkage such that a one-man crew could transfer control of the brake from one end to the other without risking its release. This would not be possible with a conventionally rigged single-truck car. Some sort of sway-bar, or its equivalent, would have been required to replace the usual two 'floating levers' and an equalising mechanism to ensure that near equal brake forces were applied to the brake beams and blocks. [57]

THE RAWORTH CONTROLLER TYPES:

The first Raworth regenerative controllers – as applied to Southport's demi-car 21, were very much of the nautical style   like the Chadburn's Ship's Telegraph variety. [58] The Westinghouse 'T1R' controller seems to have been the only controller that Company made for Raworth. It was very tall with a shroud around the base of the handle and 9 series, 6 parallel, and only one brake position – lining up with Motorman Cuff's (of Rawtenstall's car 14) description referring to an emergency notch. [59]

The extensive collection of early Rawtenstall tramcar photographs in the Rawtenstall Library shows the trams all to have had British Westinghouse 'T' type controllers.[60] The *Tramway & Railway World* of 6th May, 1909 details the electrical equipment of the first 16 cars in the Rawtenstall fleet as being Westinghouse R2 controllers and Westinghouse Type 220 motors.

There is no doubt that most, if not all, early items of Raworth's electrical equipment were made by Brush, but as time passed other manufacturers became involved. Firstly, this was indirectly by Raworth's modifications to and rewinding of, existing series traction motors to give them either the shunt or compound characteristics he needed, and later in respect of control gear.

It seems likely that the limit of regenerative development reached by Brush regarding their control gear was their H.G-2 controller around 1902-04. By this time it is suspected that their enthusiasm for prospects of significant orders was waning and Raworth was having to seek his controllers in particular elsewhere.

The well known illustration of a Raworth controller, thought to be the R6 (published in 1907 and later), has a number of features which point to an origin other than Brush when compared to contemporary photographs of Brush units (notably their H-2 and H.D-2 types). An obvious difference is the form of the magnet blow-out system. Brush used what appear to be small, individual, encapsulated coils, probably pancake wound, between the controller fingers and drum segments, whilst the Raworth model had a common single coil and a large pole-piece spanning all fingers, with refractory arc shields protruding between them and continuing amongst the drum segments. The arrangement resembled GE's B13 controller, an original design of American General Electric in Schenectady; BT-H was their UK branch supplying them here.

Comparing the two, other similarities emerge in their features:

* The pole-piece support hinge and its securing by a bolt with its own hinged handle.
* Reverse and power-brake drums being separate but co-axial
* A form of Geneva drive of power-brake drum from main drum
* The shape of the reverse key and its engagement with the top plate of the controller
* The main handle appearing removable (though may be fixed), with no form of entrapment when away from the OFF position
* The position of the terminal board

The most obvious differences were the 'hexagonal' front to the Raworth controller, while the B13 was more of a 'D' shape in plan view and the profile of the main handle, which is different to the GE/BT-H. version. By the time Raworth was looking away from Brush for equipment sourcing the B13 controller, it had been made largely obsolescent by the GE/BT-H. B18 which was widely used by new customers in the UK from about 1901 onwards. Another contender for a design based on the B13 was one marketed by Bruce Peebles and Company of Edinburgh, but probably built for them by Ganz of Budapest. Again the same similarities prevail, but their main handle is more akin to the BT-H. design than that of Raworth; again the shape of the frame in plan view is the classic 'D' shape of the B13, and not hexagonal.

It is possible to speculate that Raworth persuaded another GE licensee, possible French Thomson-Houston, to make the controller illustrated for him. The design incorporated a slipping segment which can be seen in the picture. If Raworth did use a French manufacturer it could in part explain how Bacqueyrisse came to copy the slipping segment idea for his regenerative cars in Paris in the 1927-30 period.

More or less concurrent with Raworth Traction Patents' down-sizing move from London to Manchester, WC Moore, a member of Raworth's team, left to join British Westinghouse at Trafford Park. It appears he worked in their traction department on multiple-unit systems for electric trains. There is also evidence, from two sources, that Raworth bought from British Westinghouse.

1. Payments to British Westinghouse appear in Raworth Traction Patents accounts.
2. British Westinghouse designed and appear to have produced tramcar controllers designated as "...arranged for use on Raworth's Regenerative System – Motors are compound wound."

Only one such design was done by British Westinghouse and it was given their type number T1R. It was based mechanically on their quite widely used type T1C, which would mean it had an oval frame – similar in plan view to surviving T2Cs in the UK and T1Fs in Australia. Therefore the T1R cannot have been the hexagonal frame Raworth R6 as shown.

The R6 had three notches of rheostatic brake anti-clockwise from OFF on the main handle, but the T1R had only one rheostatic brake notch anti-clockwise from OFF. This is likely to have been an emergency short-circuit brake, without graduated stages. The notches of the T1R were as listed opposite:
* 1 brake
* OFF
* 9 series
* 6 parallel

Regeneration was obtained by downward notching the main handle over the parallel and series ranges. It is unlikely there would be provision for more than three

RAWORTH'S CONTROLLER. R6.

40

positions of the reverse handle (Forward; OFF; Reverse) as this series of British-Westinghouse controllers had the handle movement entrapped by the profile of the cast top cover. It is likely that the T1R had a slipping segment and power notches similar to the Raworth R6.

While it is not possible to be precise about dating the T1R design from the limited information to hand, the order in which British Westinghouse diagram numbers were issued for their tramcar controllers may give some clue. The T1R comes just after the T1G for Glasgow Corporation and immediately before the T2C for London County Council, their respective numbers being 111059, 111061 and 111062.

It seems likely that the T1R controller would be used on the Rawtenstall fleet of 1909 as British-Westinghouse obtained the composite contract for the tramways' electrification. The T1R was the only one in their range providing for regeneration, a feature which Rawtenstall's specification required.

EQUIPPING EXISTING STANDARD CARS WITH RAWORTH'S REGENERATIVE EQUIPMENT:

This was usually carried out at the car works of the various systems with equipment forwarded to them by the Raworth's Traction Patents Ltd

The demi-cars had represented only a minor, although historically interesting, innovation in the transport world. It was the provision of sets of regenerative equipment for the conversion to regenerative control of existing standard trams of the day that constituted the major portion of the activities of Raworth's firm. This was usually carried out at the car works of the various systems with equipment forwarded to them by the Raworth's Traction Patents Ltd. This resulted in about three times as many working cars of various fleets having their standard motors re-wound with compound fields of Raworth design as there was no 'state-of-the-art' regenerative motor on the market with the windings required for the Raworth system. Brush, of Loughborough were known to rewind motors of their own design to comply with the Raworth specification for the few demi-cars that they actually built. The majority of these were essentially experimental investments by the various systems but the Birmingham & Midland Tramways Company progressed to partial fleet conversion in terms of 40 sets of JSRaworth's equipment on account of the already remarkable power saving demonstrated as early as 1904. As has already been mentioned, the unlucky Rawtenstall Corporation equipped its entire fleet of new double-decker, top-covered cars with regenerative equipment.

The City of Birmingham Tramways Company and the Birmingham & Midland Tramways Company, being subsidiaries of the BET combine, shared these 40 regenerative braking cars in terms of 14:26 vehicles numbered respectively 243-256 to the former and 25-50 to the latter. They were all Brush-built on Lycett & Conaty 8ft 6in wheelbase trucks with Brush 1002B, 33 hp motors. They had open-topped unvestibuled bodies, being built in the 1904-05 period. They were, however, short-lived as regenerative braked cars, some lasting scarcely a year as such, having their regenerative equipment removed in 1906 onwards, when the cars were re-equipped with standard control right away on account of the Birmingham Corporation veto. [61] The late Stanley J Webb has written, however, in his *Black Country Tramways: Vol I*; *1870-1912*: page 204 that, 'the use of the (regenerative) must cease, and all the cars were modified to normal equipment'. It is not known exactly when the modification was carried out, but it was probably round about 1908-9.

Mr Peter Jacques, in recent correspondence with the author, writes:

'Birmingham Corporation were most suspicious about the company's regenerative operations, not least because in response to enquiries the company gave very guarded, and in some cases, evasive replies. Consequently the Corporation was not prepared to consent to company operations on the regenerative system on their lines. The City Engineer pointed out that the Board of Trade regulations imposed a maximum voltage of 650 and the fact that occasionally the regenerative system reached a very much higher figure than this, "will furnish another reason for not adopting it'. In the light of subsequent accidents elsewhere, it seems that Birmingham were right to take the stand it did.

The Corporation took over the Dudley Road lines within the city on 1st July 1906 after which date Birmingham & Midland Tramways regenerative cars were banned. The Coventry Road service was likewise taken over on 1st January 1907 which saw the end of regenerative operations at Yardley. It was supposed, theoretically, that Yardley regenerative cars could have continued outside the boundary but since all the service ran through to the City Centre, they obviously did not do so'.

It is reliably recorded by the Superintendent of Kyotts Lake Road Tramway Repair Works in Birmingham that their 1906 batch of 50 cars ordered from Dick Kerr & Co., Ltd. of Preston (their 226-270) had been offered a system of regenerative control but that their Manager, Alfred Baker, had advised against it. [62] Was this possibly of the Johnson-Lundell variety?

Yardley Power Station:

Yardley Power Station, in Birmingham, is of historical note from several aspects. It was the power station for the regenerative car Yardley route in the First Regenerative Braking Phase of JS Raworth. So many regenerative cars were on this route that breakdowns became more than a common occurrence and regenerative overvoltage in the lines became such as to quite overcome the bus-bar pressure in the power station. The polarity of the generators there was reversed into a closed-circuit, short-circuit with the regenerating cars.

This led to a massive over-speeding of the main generators, with disruption of the rotating parts of the armature spiders hurtling through the powerhouse and switchboard, in all directions – to the imminent danger of the engineers.

This led to a massive over-speeding of the main generators

The upshot of all this was the ultimatum from the Birmingham Corporation to the two BET companies – the City of Birmingham Tramways and the Birmingham & Midland Tramways – with 40 regenerative braking tramcars between them – that as from 1st January, 1907, they would not be allowed to run regenerative tramcars along the city streets. Not unnaturally, there developed a profound local antipathy towards regeneration which Arthur Maley cleverly conquered in 1936 as will be seen, not to mention ultimately the national veto following the dreadful Rawtenstall runaway accident that was to occur on 11th November, 1911.

Raworth Regenerative Material

– A Summary

Pre-Demi-car Regenerative Experiments:

Devonport	2

The Demi-cars:

Southport	1
Gravesend & Northfleet 9 & 10	2
Glossop 8	1
Halifax 95 & 96	2
Yorkshire Woollen District 59	1
Barrow-in-Furness 13 & 14	2
Chester 13	1
Erith 15 & 16	2
Darwen 14-16	3
Plymouth 37-42	6
Greenock & Port Glasgow 40	1
Maidstone 18	1
Subtotal	23

Sets of regenerative equipment for ordinary cars already delivered:

Devonport	6
Scarborough	3
Southport (further order)	1
Bournemouth	1
Yorkshire Woollen District	8
Devonport (further order)	7

Gravesend & Northfleet	5	
Birmingham & Midland	40	
Rawtenstall	16	
Maidstone (1909)	3	(Cars 3, 5 &7 withdrawn in 1929)
Scarborough (further order)	4	
Subtotal	**94**	

Then there are these more recent discoveries from Charles C Hall's records:

Plymouth: (April 1906) ordered 6 Demi-cars at a cost of £507 each.

West Ham: placed an order for a 'Special Car' from Brush with their body, a Conaty radial truck and Raworth's regenerative control system

Birmingham, Bournemouth & Le Havre: Noted 1904 that cars fitted with Raworth's regenerative control had entered service in each case with a pattern of motor not used before, namely, Brush 1002B/Westinghouse 49B and General Electric 1000.

Southampton: Raworth had an order for regenerative control equipments (November 1906) at £100, and for converted motors at £73 each.

Sheffield: 1 car

Glasgow: 2 sets of equipment

It is always possible that the reporting of orders placed in the likes of *Tramway & Railway World* could lead to some misquoting of detail, or even late withdrawal of orders. Others again might appear with future wider research but the above figure by Ian Yearsley, augmented by Charles C Hall's recent research must be as very nearly absolute as is known today. Certainly, however, the Birmingham, Bournemouth and Plymouth orders are almost certainly duplicated and the Le Havre one would seem to be unrealistic! The Sheffield order has not been recorded before and Glasgow's merits only a brief mention in official minutes as expanded upon, below.

These give a grand total of around 120 of all such equipment including demi-cars and the early Devonport experimentals, covering some 17 different systems under the first phase of regenerative braking in tramcars. History is, however, always an incomplete study. Never can all be known about any branch or topic, nor is this proffered as being anything like the last word on the subject. There is always the possibility of yet more in all archaeology, industrial or otherwise.

Charles C Hall, one time First Assistant Engineer to the General Manager of the Sheffield Tramways, has also traced from his extensive records the following minutes:-

'26th July, 1904: The Chief Electrical Engineer is to visit Devonport to inspect regenerative equipment'

'13th September, 1904: The Chief Engineer reported and suggested tests on two cars – one single-deck and one double-decked – to be run   on two of the steepest routes: at £120 per car'

'27th September, 1904: Sub-committee recommended acceptance of Raworth's Patents tender to equip one GE58 car, with regenerative system for £165, be accepted'

'28th April, 1908: Accounts: Paid Raworth Traction Patents £5-4-6d for a car electrical fitting' [63]

Then, from discarded documents of the Glasgow Corporation Transport Department Head Office in Bath Street upon the take-over on 1stJune, 1972 by the Greater Glasgow Passenger Transport Executive, came the following Corporation Minute (see overleaf):- [64]

That the transaction has not been traced as agreed by the Town Council could leave an element of doubt as to the order being ultimately placed. However, this would have been almost out of the question as by far the most of such recommendations were simply rubber-stamped by the parent committee. [65]

Pseudo Demi-cars:

As has been seen, the Raworth demi-cars were best suited to rural services and on outlying districts, even on short, less trafficked town routes unable to substantiate the use of standard-size cars. They set a pattern of similar small cars, only with standard equipment, which continued to be built, many years after. Such systems as those of Southampton, Manchester, Warrington, York and Ayr are mentioned, and Glasgow adapted Car 92, an electrified horse car, for such purposes.

Number 92 was originally officially 'The Demi-car' but latterly simply the 'One-Man Car'. The conversion has been recorded as 1905 and the rebuild took place during the phase of JS Raworth's demi-car production of 1904-1909. The semi-official use of the demi-car name would indicate almost the only interest and acknowledgement in Glasgow of that period of advancement. The equipment was standard comprising Brill 21E 6ft 0in wheelbase truck and Witting Bros. 30 hp motors. The car was placed in service on the Finnieston Cross – Stobcross Ferry shuttle. When vestibules were fitted in 1913 the truck was exchanged for a 7ft 0in wheelbase equivalent with Westinghouse 49B motors. Interestingly, 92 was fitted with a continental type of bow collector. A new extended Fischer pattern bow was fitted in 1924. After the Finnieston service closed 92 was transferred to the Paisley Cross – Abbotsinch route until that was closed in 1933. The car was scrapped in December 1938 after some years of disuse at Renfrew Depot.

Glasgow's so-called Demi-car, 92, in Paisley's County Square.
(Photo: SJT Robertson collection)

POST WORLD WAR 1:

John S Raworth, with his demi-cars of the 1903-09 period, set a pattern of British small one-man cars for services that could not sustain anything larger. They were never popular with the travelling public, possibly (as today) because they had to queue up in the rain to gain entrance, one at a time!

Following the Fist World War, not only did a few of the remaining Raworth demi-cars receive a new lease of life (albeit by then stripped of their regenerative braking equipments), but another generation of one-man cars was found necessary. The trend was short lived and none of them was fitted with regenerative machinery. They are seen to be outwith the scope of this book.

However, worthy of mention is a further pseudo-demi-car placed in service in York, in 1925. There was a need for running a short 3-mile unremunerative service economically and an order went to the English Electric Company of Preston for a single-deck, small, one-man car. The Company still had the drawings for Raworth's last pair of demi-cars of 1908 and 1909, respectively, the Greenock and Maidstone examples constructed by the United Electric Car Co., Ltd that had been taken over by English Electric. Hence it was a simple matter to run-off a further specimen. This had air and rheostatic braking, but not regenerative braking.[67]

York Corporation Tramways One Man Car 37 of 1925. Note the close resemblance to the Maidstone and Greenock Raworth demi-cars. This car was not fitted with regenerative braking itself but is obviously using the same coachwork as the genuine Raworth demi-cars. This was illustrated in *'Tramway & Railway World':1925-Nov.12: LVIII.*

REFERENCES: CHAPTER 2

1. Tramway & Light Railway World, 1902
2. Duffy, M.G.: Electric Railways 1880-1990 History of Technology Series No.31: 2003: p135
3. Ibid: 347
4. Claydon, G.B.: Strathclyde Tram – A Way Forward: Scottish Transport: 1996: 511: 55
5. Strathclyde Regional Archives: D-TC: Vol. 46: Tramway Prints, Town Clerk's Office – Vol. 2: pp416-418: Glasgow Corporation Tramways – Opinion of Town Clerk on Working of Tramways by Corporation (p 3): M.S.S. Lamb: S.R.A.: MP-21-677 October 1889.
6. Marwick, J.D.: Glasgow – The Water Supply of the City, etc.: 1901: Glasgow: Ch. VI 1859-1876: Lamb: S.R.A.: MP-21-677 October 1889.
7. S.R.A. (Strathclyde Regional Archives: Tramway Prints – Vol. 2 (pp 204-214): The First Schedule to the Articles of Association of Glasgow Tramway & Omnibus Co.,Ltd.: Agreement of 15 May, 1871
8. Colquhoun, T.: My Reminiscences of Glasgow Town Council: 1904: Glasgow: from a news-cutting in a scrap-book in the Mitchell Library, Glasgow: Ref No. C232548-G-352 of 3-4-1905
9. The Electrician: 1903-September 18th.: p885
10. Yearsley, I.A.: Raworth's Regenerative Demi-cars: Tramway Review: 1973-Winter:10:76:104
11. Ibid: The Electrician: 1903: p 885
12. Ibid: Yearsley: 1973 –Winter: 10:76:106
13. Ibid: p 107
14. Light Railway & Tramway Journal: 1903, Jan 9th: IX: 35-36
15. Hall, C.C.: Glossop Tramways: Tramway Review: 1958: Vol.3: Issue 24: pp 171-182
16. Ibid: Yearsley: 1974-Spring: 10:77:131
17. Ibid: Yearsley
18. Ibid: Yearsley: 1974-Spring: 10:77:132
19. Marsden, B.M.: Glossop Tramways: (undated): Front cover
20. Markham, J.D.: personal correspondence: 30-9-2002
21. Ibid: Yearsley: 1974-Summer: 10: 78: 179
22. Tramway Review: 1973-Winter: Front cover
23. Gillham, J.C.: A History of Halifax Corporation Tramways: Tramway Review: 1967-Autumn: 7: 51: 65
24. Ibid: Yearsley: Tramway Review: 1973-Winter: 10: 76: 100-108. Also T.R. 1974-Spring:10: 77: 131-140 and T.R. 1974-Summer: 10: 78: 179-183
25. Pickles, W.: The Tramways of Dewsbury and Wakefield: 1980: Light Railway Transport League: pp82-84
26. Ibid: Yearsley: 1974-Spring: 10:77:131-140
27. Agnew, W.A.: The Electric Tramcar Handbook: 1909: 5th Edn.: pp 68-73
28. Ibid: Yearsley: Tramway Review: 1973-Winter: 10: 76: 102 – illustration
29. Ibid: Pickles, W.
30. Ibid: Yearsley: 1974-Spring: 10:77:132
31. Cormack, I.L.: Seventy Five Years on Wheels – The History of Public Transport in Barrow-in-Furness – 1885-1960: p 18
32. Ibid: Yearsley: 1974-Summer: 10:78:180
33. Ibid: p 181
34. Tramway Review: 1957: Issues 22 & 23 (combined): p 134
35. Ibid: Yearsley: 1974-Summer: 10:78:180
36. "Southeastern": The Tramways of Woolwich & South-East London
37. Ibid: Yearsley: 1974-Summer: 10:78:180
38. Ibid: Yearsley: 1974-Spring: 10:77:135
39. McAulay, H.N.: Personal Communication of 28-11-01 with Ian G McM Stewart, Newton Mearns: Regenerative Tramcars in England – Darwen Corporation Tramways: p 8
40. Ibid: Yearsley: p 180
41. Ibid: Yearsley: 1974-Spring: 10:77:135
42. The Tramways of Southwest England: L.R.T.A (Networks Edn.) pp 26-27 & 47
43. Ibid: Yearsley: 1974-Summer: 10:78:181
44. Ibid: McAulay: p 14 (derived from "The Tramways of Southwest England: Light Rail Transit Association).
45. Rush, R.W.: Tramways of Accrington 1886-1932: 1961: Light Railway Transport League: pp 27, 28 & 79
46. Markham, J.D.: Personal communication: 30-9-2002 on "Rawtenstall Tramcars & Other Matters"
47. Cormack, I.L.: The Tramways of Greenock, Gourock & Port Glasgow: 1975: Glasgow S.T.M.S.: p 31
48. Leach, A.: (County Librarian – Bute, Arran & Cumbraes): Rothesay Tramways – A Brief History, 1882-1936: Local production: 49 Ibid: p 30
50. Ibid: Cormack, I.L.: The Rothesay Tramways Company 1879-1949: 1986: Glasgow S.T.M.S.: p 38
51. Ibid: Cormack, I.L.: pp 54-55
52. Ibid: Yearsley: 1974-Summer: 10:78:181
53. Brotchie, A.W.: Personal communication
54. Tramway Review: 2001-Winter: No.188: p 141
55. "Invicta": The Tramways of Kent: 1971: L.R.T.L.: Vol. 1
56. Ibid: Yearsley: 1974-Summer: 10:78:181
57. Markham, J.D.: Personal communication: 1-10-2002
58. The Electrician: 1903-September 18: p 885 A 'One-Man' Electric Tramcar
59. Cuff, F.: Driver of Run-away Car: Board of Trade Official Accident Report on Rawtenstall Accident of 11-11-1911 by Lieut. Colonel E.Druitt: p2
60. Markham, J.D.: Personal communication: 30-9-2002
61. The Tramways of the West Midlands (Networks Edn.): L.R.T.A.: pp 59 & 62-63
62. Lawson, P.W.: Birmingham Corporation Tramway Rolling Stock: Part II – The Brill Cars: Modern Tramway: 1970-April:
63. Hall, C.C. (Sheffield): Personal communication: 10 July 2003
64. Longworth, B: Personal communication
65. Ibid
66. Ibid: Hall, C.C.: Personal communication 22-10-2004
67. Ibid: Hall, C.C.: Personal communication, and, The Tramway & Railway World: 1925 – November 12: Vol. LVIII: 272 et seq.

ROSSENDALE TRAM SMASH.

CARS COLLIDE ON "BAXENDEN BROW."

SIXTEEN PERSONS INJURED.

PASSENGERS' ALARMING EXPERIENCES.

15.11.1911

DRIVER'S THRILLING STORY.

An alarming tramcar accident occurred on the steep gradient well-known locally as "Baxenden Brow," on the Accrington branch of the tramway system between Bacup and Accrington, late on Saturday night, two cars belonging to Rawtenstall Corporation, who have a running agreement over the route, colliding, as a result of which sixteen people were more or less injured.

The section on which the accident took place, is a long and in places steep gradient from St. John's Church, Baxenden, to the stopping place at Accrington, but the actual scene of the mishap is perhaps the steepest portion, and is situated midway. The 10-35 car from Haslingden to Accrington came along the track as usual, and duly reached the loop near the Alma Inn, where the driver waited for an up car to pass. On re-starting the car began to gather speed, and quickly got out of control. The driver, Francis Cuff, of Rawtenstall, did his utmost to check the speed, but without success. The car dashed along at a speed subsequently described as terrific. It rocked and swayed from side to side in an alarming fashion, and the passengers became terrified. Inspector Slattery, who was on the car at the time, rushed to the rear and applied the drag brake, but this failed to have an appreciable effect. Gaining momentum, the car proceeded several hundred yards, retaining its position on the track. _____ this place is a single line

Waiting in a loop at the penny stage, just above the Victoria Hotel, was a car going towards Rawtenstall. On reaching the loop the runaway car, with its affrighted passengers, jumped the points, and the buffer crashed into the off-side of the waiting vehicle with resounding impact. The force of the collision lifted the waiting tramcar off the rails and deposited it in the gutter at the side of the road. The runaway car, its progress deflected and its velocity considerably checked, ran a few yards and came to a standstill within a few inches of a tramway pole.

Both cars were wrecked. The top of the runaway car was completely torn away and thrown across the roadway. Glass and splintered woodwork flew in every direction. The passengers were hurled from their seats, and the darkness into which the cars were plunged added to the confusion. Apprised by the report of the collision and the crash of falling glass, residents rushed to the scene, some of them scantily clad, and eager hands were quickly extended to the dazed passengers.

Houses in the neighbourhood were thrown open for the reception of the injured passengers, whilst one well-known resident telephoned promptly to the police office for police and surgical aid. Mr. Hindle, Beech Willas, Haslingden, who was on his way home at the time of the occurrence, rendered all possible assistance with his motor, his car bring-ing up the police and later conveying the injured

Contemporary newscutting concerning the Rawtenstall accident which had such far-reaching consequences. *(Courtesy Rawtenstall District Library)*

47

REGENERATIVE BRAKING CONDEMNED

The 1911 Board of Trade Report on the runaway incident and collision in Accrington starts off benignly enough but the sting comes in its tail. It was the same Col Druitt who reported on the Halifax runaway in 1904. By now, he was Lt Col Druitt and one wonders whether he recalled the part played there by the regenerative control. Whatever the actual reason really was, his *Rawtenstall Report* proved to be a landmark in the history of the British Regenerative Baking tramcar, effectively delaying further development for 20 years. Because of its significance, it is reproduced in full on the following pages. By good fortune the original newspaper report of the day has survived in the Rawtenstall Library and we are grateful to them for permission to reproduce a portion of it, on the previous page.

This view of Rawtenstall No. 3 shows several points of interest; the Raworth controller, the track brake wheel on the handbrake column and, just discernable, the track brake shoe between the axles. The fashions date the picture around 1910. *(Courtesy Rawtenstall District Library)*

RAWTENSTALL CORPORATION TRAMWAYS

RAILWAY DEPARTMENT,
Board of Trade,
8, Richmond Terrace,
Whitehall, London, S.W.,
22nd November, 1911.

Sir,

I have the honour to report for the information of the Board of Trade, in compliance with your Order of the 13th November, the result of my enquiry into the circumstances under which a collision occurred between two cars belonging to the Rawtenstall Corporation, in Manchester Road, Accrington.

In this case as car No. 14 was coming down Manchester Road between Sunnyside loop just beyond Laund Road and the next crossing loop at Harcourt Road, it got out of control of the driver and, jumping the facing points of the latter loop, came into collision with car No. 11, which was waiting in the loop for car No. 14 to pass.

There were 16 passengers on No. 14 car and 29 on No. 11 car, and 20, including the conductor of No. 11 car, were injured.

The seating capacity of each car was 22 inside and 29 outside.

Car No. 14 was a four-wheeled double-decked car with a top deck cover, 6 ft. 6 in. wheelbase, weight when empty 11½ tons, including 22 cwt. for the top cover.

It was fitted with Raworth's Regenerative Control System, and had in addition a slipper brake on each side of the car applied by a hand wheel and wheel brakes applied in the ordinary way by a revolving handle.

Car No. 11 was similar in all respects.

The accident occurred at 11 p.m. on a frosty night, and the rails were in a very greasy condition.

Description

The tramway line in the part of Manchester Road, Accrington, concerned in this case is a single one with passing places; gauge, 4ft. Sunnyside loop, where the car No. 14 came last to a stand, is on a gradient of 1 in 20.3 falling for the direction in which the car was going, and the distance between the trailing points of this loop and the facing points of the next loop by Harcourt Road, near the Victoria Hotel, is 356 yards.

The falling gradients of this piece of track are as follows, viz.:-

1 in 23.2	for a distance of	15	yards
1 in 20	do.	100	do.
1 in 17.3	do.	136	do.
1 in 24.3	do.	12	do.
1 in 29.2	do.	80	do.
1 in 45.2	do.	13	do.
	Total … …	356	yards

The loop at Harcourt Road, near the Victoria Hotel, is on the 1 in 45.2 gradient. No. 11 car was waiting just inside the loop.

The radius of the points of the turn-out is 200 feet.

Conclusion

The circumstances attending this collision were fully explained in the evidence of the motorman on car No. 14, F Cuff, and of Mr CLE Stewart, the electrical engineer to the Rawtenstall Corporation and also tramway manager.

The car had been in service all day, and was taken over by driver Cuff at 1.10 pm and he drove it until the mishap, which occurred at about 11 pm.

Cuff states that the car worked all right and he had taken it over the same route five times that day, and on the final trip everything went as usual until he stopped in what is known as the Sunnyside passing place. When he started he was on a falling gradient of about 1 in 20, and he states he fed up the controller handle normally until he reached the 13th notch when the motors would be in full parallel and the car should have gone at about 10 miles an hour. But as he had cleared the loop points the car shot away at high speed, and although he dropped sand all the time, and after first trying the electric brake, and lastly the wheel brakes and slipper brakes, he was unable to regain control of the car, which after running 356 yards on a falling gradient varying from 1 in 17 to 1 in 29, reached the facing points of the next loop and owing to its high speed jumped them, with the result that the right hand corner of the car collided with the right hand corner of car No. 11 which was waiting in the loop for car No. 14 before proceeding up the hill.

Both cars were considerably damaged at the parts mentioned, and the top deck cover of No. 14 was torn off and fell in the roadway. Car No. 11 was derailed and driven back a short distance. Car No. 14 ran on some little distance beyond No. 11, crossing to the right hand side of the road and came to a stand against an electric light standard in the footpath.

As soon as car No. 14 began to exceed the normal speed, inspector Slattery, of the Accrington Corporation Tramways, who was in the car, getting no response from the driver when he signalled to him, applied the mechanical slipper brake, but by the time he got it hard on the speed of the car was evidently too high for any braking effect to result.

Inspector Slattery also states that when he noticed the car was going too quickly he missed the usual humming sound made by the motors on these cars when regenerating.

The failure of the braking effect resulting from the motors of the car ceasing to regenerate current and transmit it back into the trolley wire is explained in the evidence of Mr CLE Stewart, the electrical engineer and tramways manager of the Rawtenstall Corporation, the cause being the fusing of the shunt coil of No. 1 motor, that is the motor which would take the current from the trolley wire first.

The result of this fusing of the shunt coil was stated by Mr Stewart to be that the car instead of being driven steadily at a speed of about 10 miles an hour was driven forward by a powerful current, and as the gradient was rather a steep falling one, the car gained at once a very high speed.

Driver Cuff states he thought the car was skidding on the greasy rails, so at first tried to steady it by using sand only, and he says he tried this for quite half the time between the car shooting forwards and the collision, before moving his controller handle back to the 6th notch and finally to the emergency or full braking notch. It was impossible for driver Cuff to even suspect what had really happened, and he is not to blame in any way for the mishap.

The car appears to have been regularly examined, and the fitter, A Thompson, whose duty it is to inspect the motor equipment, states that he had examined the motor in question on the previous day, and that then there was no sign of fusing in connection with the shunt coil.

It appears that even with the most careful examination of a car fitted with this system of control before being put into service, a similar failure might occur, and so to prevent a repetition of a runaway either cars so fitted should not be run on any but quite level routes, or else some further precautions must be taken to prevent the fusing or breaking of the shunt coils.

I would suggest therefore that if the latter course be decided on the Rawtenstall Corporation be asked to send their proposals in the matter as soon as possible to the Board of Trade.

Personally, I do not consider cars fitted with regenerative control are so suitable as ordinary cars for steep gradients, as it is not possible to coast down a gradient with the former by means of the application of the slipper brake and wheel brakes, as the current cannot be cut off by the controller handle being brought into the off position, until all the electric braking effect has been brought into play, and the car thereby brought almost to a standstill.

It will be seen from the time sheet of driver Cuff's duties for the week, appended to this report, that he had worked very long hours on the day of the accident, though he had only worked 60 hours in the week ending that day, for which number of hours he is guaranteed payment.

It will be further observed from the explanation with the time sheet that the extra tours on the 11th were due to another man failing to take up his proper turn of duty. But the short period of rest of only 6 hours between going off duty at 11.20 pm after 10 hours' duty and coming on duty again at 5.30 am is quite insufficient, and steps should be taken to prevent a repetition of such hours.

I have, &c.,
E Druitt,
Lieut.-Colonel.

> Here is John Markham's commentary on Lieut-Colonel Druitt's report, bringing his own knowledge and professional experience to bear, and it asks some pertinent questions.

There are a number of significant factors which do not seem to have been taken into account

    What does not appear to have been established is the true extent of all problems that may have existed. There are a number of significant factors which do not seem to have been taken into account, and the present day practice of risk assessment does not appear to have been applied.
    First of all, the Board of Trade rules for the equipment of tramcars stated that there should have been 'an emergency switch provided for the driver by which he may cut off the current in the event of the failure of the controller'. No reference is

made to this in the Accident Report, nor to the lack of its use by Driver Cuff or Inspector Slattery. This could point to a possible weakness in the training of tramway staff by Rawtenstall and Accrington Corporations.

Furthermore, Lieut-Col. Druitt does not appear to have appreciated the difference between a shunt coil and a shunt winding. He uses the term 'coil' when 'winding' is implied by the text. There would be four coils in series, one per pole, forming the winding. The failure would have been at one of the coil connections, of which there would be several in series per motor. I have doubts that all of them (if any) could be seen adequately by the form of inspection that fitter Thompson could have undertaken. He could truthfully say that he had not seen a fault when he examined the motor. However, a question such as 'Can you say for sure that all shunt coil connections were in good order?' would have been much more difficult for him to answer.

Is it possible that the far higher than average number of breakdowns, requiring recovery, involving regenerative Birmingham & Midland company cars, when running on the Birmingham Corporation tracks in 1905-6 suggests faults of the shunt coils (or their connections) were a regular and unwelcome feature? Was the possible danger of accidents arising from this cause identified? (The importance of analysing near-miss incidents cannot be over-emphasised). Due to their unreliability, Birmingham Corporation forbade the use of these cars within the City from 1st January, 1907.

It is interesting that the first sign of the incident took place just as car 14 entered the single line. This makes me question if a sudden jerk to the motor concerned – perhaps caused by the negotiation of the pointwork – contributed to the final break of the shunt field connection. It must be a possibility, though I accept that the connection must already have been weak: a classic case of fatigue failure. Were any shunt coils loose on their poles?

It is most surprising that the 'powerful current' mentioned in the report as 'driving the car forward at high speed' did not trip the car's automatic circuit breakers. The current concerned would be likely to be several times (say 2-4) greater than that normally taken by the car, perhaps of the order of 300-500 amps. Normal maximum could be expected to be in the range 200-250 amps. It is quite possible that previous similar incidents elsewhere were protected by the tripping of one or both of the tram's circuit breakers before any major change in speed could ensue, thus averting catastrophe. Both Lieut-Col. Druitt and Mr Trotter are silent on this fundamental matter. One wonders why.....

Druitt was clearly a humane person, perhaps covertly anxious that neither Cuff nor Thompson should be sacked over the incident. For this he deserves merit. Certainly he probed the incident sufficiently deeply to establish that he had a route to put the cause simply down to a failure of equipment that would then be well out of warranty.

Lieut-Col. Druitt's opinion about the unsuitability of regen equipment for hilly routes is, I feel, unfair and far too sweeping. If the Johnson-Lundell Company had still been active they could have challenged his statement. With their system it was always possible for the driver to return his controller to OFF without causing the car to decelerate, simply by not selecting the regen. braking function. The Raworth challenge to this would be that their system, where the driver did not have this option, was bound to achieve greater use of the regenerative feature as it had to be used every time. They may even have applied this argument during the promotional phase of regenerative equipment in previous years.

Then his remarks about the unsuitability of Raworth regeneration on down gradients could also be challenged. The car could coast down a hill under the control of the slipper brake by simply allowing the car to start by gravity on the falling gradient and leaving the controller in the OFF position. In this particular case it would have been with no loss to Rawtenstall. The saving benefit from regeneration, where the incident occurred, mainly accrued to Accrington, the car being on their electrical system!

The remarks in the appendix by Mr Trotter are of particular interest and significance. His first sentence could give every purchaser of Raworth equipment an immediate claim against Raworth Traction Patents, even though trading by that company was, by this time, somewhat in the doldrums. Sensing a liability which it might well not be able to meet may well have prompted its liquidation if his remarks did indeed apply to all units. Was it a design defect, or one of quality control? The question does not appear to have been investigated. His comments in relation to corrosion suggests that he thought that any chemical flux used when making a

His first sentence could give every purchaser of Raworth equipment an immediate claim against Raworth Traction Patents

soldered connection between the coil termination and its winding wire may not have been thoroughly cleaned off after the joint had cooled. Evidence of similar corrosion can often be seen today on copper pipework which has soldered joints in the form of green patina near the joint.

Mr Trotter's advice on the termination fine wire coils identifies the weak point of the assembly, but it should be remembered that the wire of the day used for such coils would probably be insulated with wrappings of cotton or silk. Little or no abrasion of the wire end to remove a synthetic enamel would be needed. With present day winding wires, with their flexible enamel insulations, the junction between winding wire and termination wire, if made in the manner he describes, would still be liable to failure near the start of the section from which insulation has been removed. A multiple strand insulation technique is advised.

The remarks by Druitt and Trotter do not appear to have deterred the engineers who prompted the second generation of regenerative equipments of the 1930s. Whilst the details of the designs differed, the basic principles did not. Halifax, one of the users, had the hilliest system in the nation.

The remarks by Druitt and Trotter do not appear to have deterred the engineers who prompted the second generation of regenerative equipments of the 1930s.

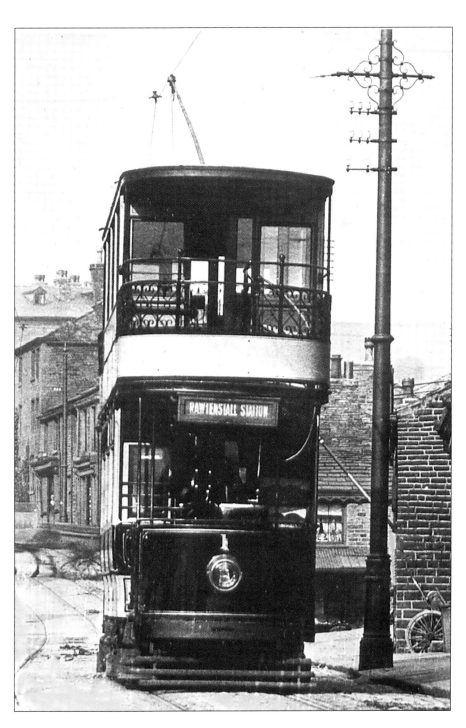

Rawtenstall No. 1 heads for the Station, having just descended one of the gradients that were characteristic of the system.
(Photo: SITA collection)

THE GENERAL HISTORY OF REGENERATIVE BRAKING IN ROAD TRANSPORT VEHICLES AND ITS INFLUENCE ON TRAMCARS

Although this book deals mainly with the British Regenerative Braking *Tramcar*, the use of this technology during the period between the first phase and second phase in trams formed a transition in development that made the second phase a more practical proposition. Struan Robertson explains.

Regenerative Braking has been applied both to electric tramcars and to railless electric road vehicles. Therefore, in the delineation of its history from a road vehicle aspect, there is inevitably a certain dodging about from one system to another, unavoidable when attempting to deal with it chronologically. The Phase One Regenerative Braking in tramcars has already been covered although some information relating to this will re-emerge to highlight parallel and inter-related developments.

The early experimenters in practical electricity were aware of the characteristics of dc electric machines to perform both as motors and as generators. This depended on whether they were fed electric power or were driven mechanically while coasting. Always mindful of this, transport electrical engineers have yearned after the ability to harness both these characteristics in the interests of economy. This book is all about the practical aspects of the outcome of this yearning.

Recent research by John Markham has indicated that probably the first potential application of regenerative braking in this country occurred in 1889 (his Chapter 'Origin of Tramway Regenerative Braking' refers). It had been preceded in America, source of our big names in early electrical manufacturing.

THE BIRTH OF THE ELECTRIC ROAD CAR

French engineers were ahead of the world in the introduction of electrical automobiles. There were some 500 on their roads by the mid-1890s. Perhaps the most interesting of these, because it was the first to use regenerative braking, was an electrical coupé displayed in Paris by MA Darracq in 1897.[1]

The 'Red Flag Act' stifled development in Britain but following its repeal, quite suddenly, electrical vehicles started to emerge and at last England began to catch up. One of the firms involved in the early building of electric automobiles with regenerative braking in this country was the Johnson-Lundell Electric Traction Company of Southall. It also applied regenerative braking to its other products.[2]

Their road cars were of two types. One battery powered all-electric version had two motors mounted on the rear axle, while the other generated its electricity on board by means of a combined oil engine and dynamo feeding the electric motors. John Markham rightly queries the Johnson-Lundell claim made in Rankin Kennedy's

'Electric Installations' back in 1909, pages 56 & 57, where an illustration of Johnson-Lundell's electric road car shows a totally inadequate provision of leads entering the two motors to have had regenerative braking equipment, despite its being positively asserted on page 57, paragraph 2.

The year 1896 saw the introduction of electric taxis in New York, London following in the next year. The regenerative braking allowed their range to be increased by 5-15% and also offered battery re-charging.

It was the extensive development work carried out by Johnson-Lundell for road vehicles that enabled them to put the first regenerative braking tramcar on the road in this country

It was the extensive development work carried out by Johnson-Lundell for road vehicles that enabled them to put the first regenerative braking tramcar on the road in this country. It was demonstrated to the media at their premises in London, followed up in 1902 with a lengthy period of assessment on the recently electrified Newcastle-upon-Tyne system under the independent eye of Paschal, a consulting engineer of the day, who oversaw the results of the various tests.[3]

THE ONSET OF PHASE ONE OF REGENERATIVE BRAKING IN TRAMCARS

Although primarily developed for battery-electric road automobiles, regenerative braking in tramway practice was resorted to solely in search of economy in service. Generating authorities in those days were small, strictly local, low powered and almost totally catering for domestic use. The industrial use of power commenced with tramway demands, otherwise generation was almost entirely an early morning and nocturnal demand.

Where the early electric tramways did not generate for their own use they had recourse to the existing local generating authorities which thus had to cope with an average 500Volt requirement. Not unnaturally an additional charge was levied on their rate of production, to compensate. Where local generating authorities were also the Local Authorities these power rates tended to soar disproportionately. In the final decade of the nineteenth century and early years of the twentieth century the Local Authority was not allowed by law,[4] (The 1870 Tramways Act), to run tramways within their own jurisdiction. They vented their resentment in terms of the cost of the provision of the necessary power. It was as a direct result of such policy that transport electrical engineers sought means of economising via two channels: cheaper current and smaller, cheaper, one-man-operated cars.

Access to cheaper current, where the generating authority was impervious to reasoned negotiation in those days of soaring civic socialism, could only be achieved through the harvesting and application of the regenerative characteristic of the dc traction motors of the service tramcars. Many were the engineers researching the project but two firms, only, seemed to attain viable conclusions. They were the Johnson-Lundell Electric Traction Company Ltd. of 16A Soho Square, London [5] – essentially a British branch of an American mother-company, and John Smith Raworth of Raworth's Traction Patents Ltd., Queen Anne's Gate, Westminster, of 1st October 1904.[6]

Johnson-Lundell, although a small firm, had been active in London since the early 1890s. They were forward looking, producing innovative designs of electric machinery such as electric dynamos, traction motors for railways, tramways and automobiles, of compound, regenerative type, regulating entirely by counter electromotive force (all armature resistances being disposed of). They also acted as dynamic brakes. In this capacity they returned to the line, in useful work, a portion of the moving energy of the car in descending gradients or in arresting momentum. They were also the inventors and providers of the Company's surface-contact system,[7] which had been considered, but not accepted, for the Aberdeen Tramways.[8]

Raworth, with his wide training and experience in both the mechanical and electrical branches of engineering, not to mention having a brilliant inventive genius, brought to bear his vivid intellect as a tramway man upon the most important aspect of tramway function – the economy in consumption of electricity. As with the electric locomotive, or train, of the day, the normal tramcar was extremely wasteful of that commodity in consuming heavily in its resistor banks both in acceleration and in braking. In the coasting mode high values of potential regeneration were allowed to go unharnessed. Primarily Raworth's dream lay in effecting this harnessing and his two arch-enemies appear to have been the marked inadequacy of development of the traction motor and, perhaps, a failure fully to realise the extreme importance of having a receptive overhead, one with sufficient load to absorb, more than comfortably, the very high and very irregular values of regenerated current. Despite these Raworth's system worked very well initially but tended to deteriorate in function after about a year or so. It had to contend with overheating in the unventilated motors coupled with the violent jerks of acutely short radius bends and the interminable rattle over points, crossings and uneven permanent way. These factors were particularly heavy on the internal connections of motors, perhaps sometimes to their undoing.

Granted these, Raworth did produce a regenerative system at once functional, less complicated, less heavy and very much less expensive to install. It was perhaps a modicum less perfect in function than Johnson-Lundell's. These factors, together with his intimate acceptance into both the manufacturing and executive circles of British tramways, may have secured for his firm the majority of orders along the 'old boy' network, to the chagrin of the Johnson-Lundell combine.

As we have seen, an ill-advised resort to litigation over an alleged breach of patent reduced the Johnson-Lundell firm to insolvency and brought the Raworth firm to its knees. This was followed by the disastrous Rawtenstall collision (described elsewhere) and precipitated the voluntary wind-up of the JS Raworth firm and the end of the First Phase of Regenerative Braking in tramcars.

Johnson-Lundell's System Examined

Much thought had gone into the Johnson-Lundell machine. Its 35 hp motors were compound-wound, had completely laminated magnetic circuits, double commutators (one at each end of the armature) and used flat-strip conductors for both series and shunt windings. A major requirement of these motors was the ability to utilise their back-emf for braking and when descending hills. A material amount of energy was restored to the overhead in regenerative braking. The gear ratio was 69/14, and the car wheels were assumed, in reporting, to be the recognised standard of the time, of 33in diameter.[9]

Their system of regenerative control was arranged to perform as series-wound machines when accelerating and running, and as compound-wound machines when regenerating and braking. Expensive and heavy duplicated control was exerted by the motorman's direct platform controllers through an underslung secondary controller, called a 'Field Changer'. This altered the motor magnet field

Raworth's system worked very well initially but tended to deteriorate in function after about a year or so

. . . an ill-advised resort to litigation over an alleged breach of patent reduced the Johnson-Lundell firm to insolvency and brought the Raworth firm to its knees

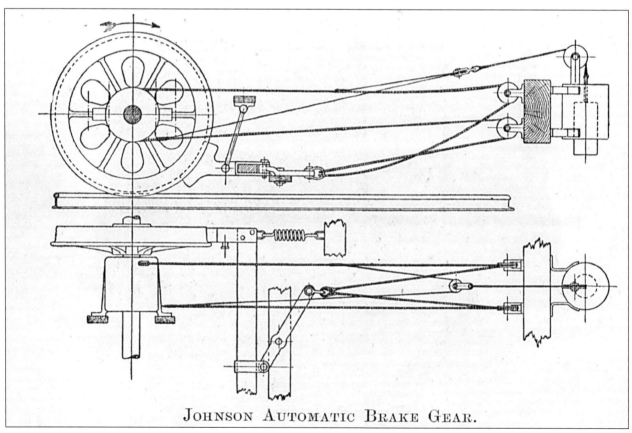

JOHNSON AUTOMATIC BRAKE GEAR.

connections from series to compound for regeneration and braking purposes by the simple pressing of a small button on the main handle of the platform controller. Credit for the development of this system goes to Gustav Lang.

Regeneration was neither capable of bringing the car to rest nor of holding it on a gradient. The Johnson-Lundell firm had invented an automatic brake gear utilising the drop of a heavy solenoid iron core causing a stout steel cable, working over a train of pulleys, to tighten up around the drum of the car axle. This pulling round of the five circumferential friction band wound round the drum of the car wheel also pulled upon the main brake-beam actuating lever, pushing the brake-shoes on to the wheel reins. In doing so it applied the brake shoes to the wheel rims – a very intricate, but ingenious system of steel cables.[10]

There is conflicting evidence as to the regenerative equipment of the original Rawtenstall sixteen double-deck single truck cars of 1909. Of most interest the highly competent transport historian and journalist Ian Yearsley has failed to find any record of Johnson-Lundell regenerative equipment throughout the country.[11] Rush first pronounces that the 'majority of Rawtenstall's tramway fleet was Johnson-Lundell equipped'[12] and then proceeds to the counter statement that half were Johnson-Lundell equipped and half Raworth equipped![13] The *Tramway & Railway World* of 6th May 1909 (page 347) states that –

'The car motors by British Westinghouse Company's No.220 pattern were wound for Raworth's regenerative control, of 42 h.p. each, and Westinghouse controllers were of the 'R2' type'

and as such determines the equipment for the initial fleet of sixteen cars. The Board of Trade's Inspector of Railways, Col Druitt, RE, in reporting, officially, the dreadful Rawtenstall crash of 1911 stated that all of the initial fleet of Rawtenstall tramcars were Raworth equipped, not Johnson-Lundell. This very great preponderance of technical opinion therefore lies with Raworth's equipment. Yearsley's failure to trace any Johnson-Lundell regenerative equipment at all in the country adds powerfully negative evidence against the use of Johnson-Lundell equipment. It can be taken that all of Rawtenstall's first sixteen cars were Raworth equipped. Subsequent research has elicited the fact that certainly one passenger car in Britain had been fitted with Johnson-Lundell regenerative equipment – the 1902 Newcastle experiment previously referred to.[14]

Certainly the Board of Trade Report on the Rawtenstall disaster and its (dubious?) conclusions referred to previously, terminated the first phase of Regenerative Braking in Tramcars in Britain, but equally certainly, not so in railless road vehicles.

No information has come to light as to which Newcastle tram was equipped with the Johnson-Lundell regen equipment. It may well have been a car like 176 descending Shields Road. *(Photo: BJ Cross collection)*

Early Developments with Regenerative Braking in Railless Road Vehicles

WA Stevens writes in the *Tramway & Railway World* of 13th November 1930 (page 283) that he had conducted experimentation in regeneration from electrical machinery for the past 25 years. This was at first on battery vehicles and then on trackless trolley vehicles. Having associated himself with trackless trolley design and construction, Stevens maintained this interest, perfecting as he went, through to the 1926 boom in British trolleybus construction, and well beyond. With the firm of Guy Motors Limited he designed more powerful motors, perfecting regenerative braking control and building up a most substantial understanding of the technology of regeneration. With this intensive interest he attended the 1930 Paris Conference, referred to later in this chapter, on regenerative braking in tramcars and was led to design and build regenerative motors for trams evolved from Guy's highly successful and efficient trolleybus motors. Their offer to co-operate with the Manchester experimentation (see below) using their own design of machines was – sadly – turned down. Very much might indeed have come of it!

The first trackless trolley in the United States is generally regarded as that at Laurel Canyon near Los Angeles and occasional individual experiments with steered trolley-fed, electrically driven buses had been made from time to time since then. All were unsuccessful until a French engineer from Lyons exhibited and ran a small trolleybus system in connection with the 1900 French Exhibition. Even this

<table>
<tr><td>

WA Stevens

The 'Stevens' in the Rees-Stevens is the same as in 'Tilling-Stevens'. He established his company in 1897, bearing the name 'WA Stevens', Manufacturer of Petrol Electric Buses from 1906. It was taken over by Thomas Tilling, not only as an addition to its portfolio of interests, but also because of their use of buses with petrol-electric transmission. The company later became Tilling-Stevens.

</td></tr>
</table>

Commercial postcard view of the French exhibition at Lyons in 1900.
(STIT collection)

was short of commercial success! However the trolleybus was accepted in France as a commercially viable system two years later.[15]

1903 saw a short but successful line of trolleybuses in Grevenbruck in Westphalia described in the *Tramway & Railway World* of 12th March 1904, shown above from a contemporary French source. By the following year many miles of trolleybus routes had sprung up on the continent but it was not until 1909 that the Railless Electric Traction Company built a short experimental line in the Hendon Depot yard of the Metropolitan Electric Tramways Company in London. Leeds and Bradford, both realising its potential, applied for, and received Parliamentary powers to construct trolleybus systems in 1910. These simultaneous projects were the first successful British trolleybus systems on 20th June 1911.[16]

This was followed almost immediately by Keighley. These earliest trolleybuses were very square-ended wooden single-deckers built by Messrs Hurst Nelson & Company Ltd., of Motherwell, essentially tramcar bodies mounted on what was virtually a motor-lorry chassis incorporating a pair of Siemens Bros motors[17]. These 20 hp motors operated at 525 Volts and 1050

The RET's 1909 demonstration trolleybus was painted in the MET livery presumably in the hope of an order. *(Geoff Lumb collection)*

rpm and were of the series-parallel, shunt field, commutating pole (interpole) pattern slung transverse to the vehicle, in line driving through double reduction worm-wheel-final chain gearing, each to one solid tyred wheel, aft, and weighed 3 tons 12 cwt. Their 26-passenger saloon floors were perched high above their solid tyred wheels approached by a thin four-stepped narrow staircase effectively barring their use to all but the most agile. They were reasonably successful, with Keighley following suit very shortly after, and Rotherham in 1912 showing improvement with lower pitched saloons approached by only three steps.[18] The design of trolleybus was already showing improvement. Yorkshire was truly the birthplace of the British trolleybus!

The Bradford trackless trolley, or trackless tram, of 1911, of the first batch of trolleybuses in Britain, coincident with Leeds. The photograph shows the port-side motor, of the two slung amidships, and the very high approach platform requiring four steps to the approaching stairway so precluding all but the most agile of passengers. The body was by Hurst Nelson & Co., Ltd., Motherwell. *(Photo: The Tramway & Railway World, courtesy CC Hall)*

Leeds Corporation trolleybus 505 at White Cross, Guiseley, in 1925, with Leeds-built tram number 333 in the background at the terminus. *(STA)*

Rotherham Corporation Tramways trackless trolley of 1912 on the turning circle at Maltby. Note: already the loading platform has been lowered in the design. This one has now only three steps to the approach staircase for loading – an improvement on the first trackless trolleys of both Bradford and Leeds of the year before. *(Photo: CC Hall Collection)*

Inevitably their further development was stalled by the advent of World War I. After this, research in regeneration did go on, but not any longer on the tramways. Electrical history seems never to have recorded precisely the date of onset of use of interpoles in traction motors. Raworth never had the pleasure of their use having ceased production of such equipment for tramcars in 1911 or so. Yet here in the same year is to be found the first British trolleybus sporting four-interpole motors. *Ergo* the onset of the interpole traction motor was 1911, by deduction! No less an authority than the Manchester Corporation Transport General Manager R Stuart Pilcher estimated its arrival as 1919.[19] Application for enlightenment for this book, from the Institute of Electrical Engineers having elicited no immediate knowledge, a research for the date was offered at £100 per hour. 1911 will do for our purposes!

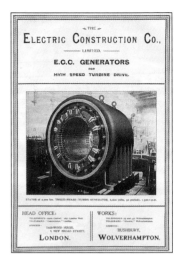

The Electric Construction Company was a huge combine with interests that ranged from producing motors to building power stations and equipping tramway systems. *(STA)*

A 1921 view of Keighley Corporation trolleybus 58 at Lidgett Hill, Oakworth, returning to town. Note the early form of current collection, in this case the German Cedes-Stoll system. *(STA)*

Keighley Corporation's No. 3 had a front-entrance body by Dodson taken from a former Cedes-Stoll vehicle. Note the unusual pivotal arrangement for the two trolleypoles whereby they are positioned one above the other, rather than side-by-side as later became standard. One advantage of this early system was that it allowed the whole assembly to rotate through 360°. *(STA)*

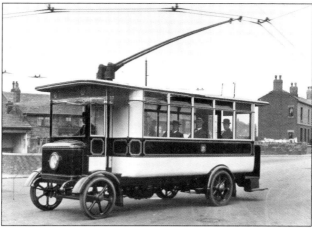

the trolleybus began to gain in popularity with bigger concerns – especially if they had their own generating plant.

The 1911 Board of Trade veto was never rescinded

THE MID-1920S TROLLEYBUS AND REGENERATION BOOM

The post-Great War period had seen gross arrears in maintenance of all aspects of transport and the tramways accumulated more than their fair share. This led to heavy programmes of deferred but highly necessary repairs. These became less and less feasible to carry out as the post-war depression set in coupled with rising costs of labour and materials and general shortages. Faced with these, smaller systems either converted to motor-buses or bowed out of business. Although the petrol bus was almost ubiquitous as the replacement for worn-out trams and tramways for operators determined to stay in business, the trolleybus began to gain in popularity with bigger concerns – especially if they had their own generating plant.

The earliest double-decker modern trolleybus with regenerative braking was the 1926 Guy six-wheeler, 33, in the Wolverhampton fleet. Just about every Guy trolleybus built between then and 1940 had regenerative equipment. Until about 1935 this was the Guy Rees-Stevens system with Rees-Roturbo traction motors, but it was thought that after this the system changed and the motors were supplied by the Electric Construction Company of Wolverhampton. The Metro-Vic and British Thomson-Houston firms were second in the field from around 1932 and their system was reported to have been quite different from the Guy Rees-Stevens type. English Electric had regenerative equipment available from about 1933 on AEC trolleybuses. GEC and Crompton-Parkinson/Allen West came later into the field, around 1936. [20]

The great majority of Guy trolleybuses used regenerative braking with single, high-powered, series-wound motors. This was the concept of Mr Stevens and great economy in current was achieved with very little trouble. Wolverhampton achieved 20% economy in running and the South Lancashire Tramways 60 seater vehicles operated with an average 1.52 units of electricity per mile.

These certainly led the way to massive installations in England. Scotland never took the trolleybus concept to mind, very much. Dundee thought about it (eventually operating two) but they were quickly sold to Halifax where they were scrapped on 31st July 1926. [21] Ultimately, in Scotland, only Glasgow took them up, very much later, to use cheap Pinkston Power Station output, as the tramcar fleet was run down. Their's was the last new trolleybus system but the vehicles were not regenerative, having series-wound motors without compounding – the cheapest option.

The 1911 Board of Trade veto was never rescinded. However, other incidents led directly to the Second Phase of Regenerative Braking in tramcars in the ten years before it emerged. These included the invention of the interpole, the improvement of ventilation of motors and adoption of roller bearings. Yet it was the very rapid rise of the trolleybus in the mid-1920s and its successful application of regenerative braking during the following years that must have led to its reconsideration in tramcars – even if only at a subconscious level.

Birmingham Corporation heralded the start of post World War 1 trolleybus redevelopment in 1921/22, becoming the first tramway operator to replace one of its tramcar routes with four-wheeled double-deck top-covered trolleybuses running on solid tyres. [22,23,24] However, it was not until 1924-25 that real and active interest began to develop in the trolleybus. At the same time Bradford Corporation placed large

Bradford Corporation's home-built 522 featured twin-steering a mock open balcony, mimicking open-balcony trams. By 1922 contemporary trams were becoming more attractive, but by that time the trolleybus still had some (considerable) way to go. *(STA)*

Compared with Bradford's 522 Leeds Corpration's No. 512 was positively hideous and looked like a tramcar which had gone badly wrong. The front axle swivelled as a complete unit and must have been heavy going for its unlucky drivers! *(STA)*

Keighley's No. 5 dates from 1924 and comprises a Brush body on a Straker-Clough chassis. Note the upper-deck still did not extend over the driver's cabin, no doubt to mitigate the heavy steering. The lighter weight bodywork is clear to see. *(STA)*

capacity six-wheeled trolleybuses in service, giving fresh impetus to that form of traffic.[25]

As previously mentioned 1926 saw Guy Motors Limited of Wolverhampton established in the field. This was the 1926-30 Trolleybus Boom and led to the supply of an overseas demand with exports to the continent and beyond – as far as Japan. 1926 may be looked upon as the year of establishment of the modern trolleybus – especially with regenerative braking. By 1927-28 there were 20 trolleybus systems, and by the following year their number had grown to twenty three. By 1929 the Hastings Tramway Company had the second largest fleet of trolleybuses in the world, having standardised on Guy six-wheelers of both double- and single-decks.[26]

London was a later starter with trolleybuses with its first fleet of 60 launched in 1931. These were the famous fleet of "Diddlers".[27] They did not incorporate regenerative braking although one (11) did receive regenerative equipment during pre-war years.[28] Later vehicles, built from 1934-1940 were regenerative.[29] Ashton-under-Lyne and Manchester bought regenerative trolleybuses from 1938-1943, Bournemouth had them from 1934 (all the pre-war Sunbeams). South Lancashire Transport's pre-1943 fleet were regenerative as were much of the pre-war fleets in Huddersfield and Hull. The bulk of the South Shields fleets, both pre- and post-war and the Newcastle pre-war fleets were all regenerative braking vehicles.[30]

Weymann's supplied the bodywork for these Wolverhampton Corporation Sunbeam MS1 and MS2 trolleybuses which dated from 1931. By this time trolleybuses were becoming more pleasing in appearance although these examples were ahead of their time when compared with examples in some other fleets. *(STA)*

On the right is a photograph taken in 2005 at the Black Country Museum of one of the Wolverhampton vehicles, originally number 78, and rescued for preservation. It will be a long term project and, sadly, the all-important motor is missing. *(Photo: John A Senior)*

So very popular had the trolleybus become from its economy of current alone, not forgetting the rapidity of starting and stopping, safety, comfort and manoeuvrability that they became a close contender for the petrol motor bus. On the continent could still be seen its counterpart, the trolley-lorry, witness the system at Tirano in North Italy, which operated in association with a nearby hydro-electric generating scheme in the southern Italian Alps. The sheer economy of operation obviously dictated its use there.

Early trolleybuses had to use the standard tramway motors of the time: heavy, cumbersome and lacking in horsepower, but there was nothing else on the market. It was Guy Motors Limited of Wolverhampton that introduced the single-traction-motor trolleybus with great success, departing from the standard tramcar practice of twin motors working in series-parallel control. This reduced the overall weight of the vehicle and offered a superior performance with higher seating capacities. The breed became even more attractive as a replacement for smaller and less remunerative tramways and such systems were progressively abandoned.

It was Guy Motors again that first introduced regenerative braking to the modern trolleybus motor using a machine providing 75 bhp, at 500Volt and 900-2,500 rpm rating which they had come to prefer over the series-wound motor and in which the shunt circuit was never interrupted. To prevent regenerated voltage rising above line voltage, when coasting, a resistor was automatically brought into circuit with the shunt field, keeping the voltage within safe limits. The company had really set the

Opening day at Twickenham of the London United trolleybus system in 1931, with two of the famous 'Diddler' AEC 663T vehicles. One of these became regenerative, heralding widespread adoption of the principle in London. (STA)

It was Guy Motors again that first introduced regenerative braking to the modern trolleybus motor

Trolleybuses were seen in early days to have a future for goods traffic in rural areas. A late example of this (using locally generated power) could be found near Tirano, North Italy. Note that automatic trolley retrievers are fitted. (Photo: Struan JT Robertson)

South Lancashire Transport took a second batch of Guy trolleybuses, this time two-axle models, but again with Rees-Turbo regenerative motors, when they replaced the last of their tramcars. One is seen here at the Crossgates works of the Leeds bodymakers Chas H Roe. *(STA)*

ball rolling with a vengeance and exported lavishly abroad. They had perfected the genre and naturally preferred it above all others.

Guy Motors Limited had researched the economy of the trolleybus versus the tramcar in their quest to sell the former and produced the following figures for the year ending 31st March, 1929. It should also be noted that tyre costs dropped by 5%-6% below that of the petrol bus due to the more gentle overall braking of the regenerative vehicle which did not apply mechanical braking until the speed had dropped to 6-8 mph.

	Tramcars	Trolleybuses
Average operating costs per car-mile	13.599d	10.968d
Maintenance and repairs per car mile	3.059d	9.365d
Fuel	1.99d	1.634d
(1p = 2.4d)		

Maintenance costs were low, and street mobility became enhanced to the point of kerbside loading. This increased passenger safety while boarding and dismounting as well as for overtaking vehicles by causing less road congestion. The Guy trolleybus of the mid-1920s, with regenerative braking, was rated at a commercial efficiency of over 90%.[31]

The trolleybus has been seen as the development of the tramcar into railless exploitation and it was perhaps inevitable that the concept of the well-tried and traditional series-parallel drive with twin motors should have been considered necessary for the developing trolleybus. So there were many examples of such before the benefits of the single traction motor with compound winding began to be appreciated.

EVENTS LEADING TO THE SECOND PHASE OF REGENERATIVE BRAKING IN TRAMCARS

The true stimulus in Britain was undoubtedly the 1930 Paris Conference that followed the Continental engineers' extensive practical experience over the handful of years leading up to it. It was in 1927-28 that both Germany and France became interested in regeneration and four cities (Nuremburg, Chemnitz – later Karl Marx Stad – Paris and Marseilles) had co-operated in drawing up a progress report on regenerative braking that was given a forum at a meeting of the International Tramways, Light Railways & Public Automobile Transport Union (IT & LR) held in Warsaw between 30th June and 2nd July 1930.[32] Sadly – and this was particularly remarked upon – only two British Tramway concerns were represented at Warsaw, these being Manchester and Glasgow. Was it a coincidence that these same two British systems would originate extensive regenerative braking research at home within a short time?

THE PARIS CONFERENCE

M. Louis Bacqueyrisse had read a paper on 'The Means and Methods for Facilitating Passenger Traffic' at the Warsaw meeting. He was the Director General of Operation and Technical Services of the Societe des Transports en Commun de la Region Parisienne (STCRP) and General Manager of Works and Technical Services of the Public Transport Company of Paris. He was the engineer responsible for the design and application of regenerative braking to the Paris Tramways. It was following the meeting in Warsaw that the President M.Frederic de Lancker of the International Tramway & Light Railway Union offered to arrange a special meeting in Paris towards the end of October 1930 to demonstrate the results obtained by the Paris company, to those interested, from the working of regenerative braking motors. This duly took place on 21st and 22nd October and turned out to be a most popular meeting. It was attended by some 200 members from many European countries – 35 from Britain, under the leadership of R Stuart Pilcher, then President of the Municipal Tramways & Transport Association (MTTA), and General Manager of Manchester Corporation Transport Department. He was accompanied by 18 General Managers of British systems, 5 lay members of civic tramways committees, GH Fletcher, Chief Engineer and Manager, Traction Motor Works, Metropolitan-Vickers Electrical Company Limited, Sheffield, WA Stevens of Guy Motors Limited

and various officers of transport departments, motor engineering firms and representatives of the *Tramway & Railway World* and the *Electric Railway, Bus & Tram* periodicals, London. All gave tacit indication of the marked interest that had been aroused in Warsaw. Mr Pilcher, incidentally, arranged for the British party to be conveyed all the way in an as yet undelivered Crossley double-decker bus that would join the Manchester fleet as 249. It was the first covered-top double-decker bus in Paris and aroused much comment![33]

M. Bacqueyrisse's paper was published in the *Tramway & Railway World*, 13th November, 1930 edition. This was followed by summaries by GH Fletcher of

M. BACQUEYRISSE,
Ingénieur en Chef des Tramways à la Compagnie Génér des Omnibus de Paris.

While in France members of the Union proceeded to Versailles to inspect the Versailles Electric Tramways and see their regenerative control system in operation. Stuart Pilcher took his MCTD staff photographer on the trip, together with other selected invitees, and naturally a visit to the famous Palace was also on the agenda, as was the obligatory photographic halt, complete with a few locals to add to the ambience. *(STA)*

Several of the famous Parisian buildings were visited, but this one was the *raison d'etre* of the visit, the offices of the Paris Transport organisation, whose name can be seen above the windows on the left. This time the local buses are included in the view though unusually Mr Pilcher is not. The party would be hearing about plans for a new tram from SOTRAMO shortly to be introduced in Versailles, and seen below. Pilcher had already recognised a problem with energy costs in Manchester, and, true to Raworth's thinking, had seen that regenerative braking could offer a possible solution. It would be some years before he had the real opportunity forced upon him with the introduction of trolleybuses in 1938. *(Courtesy HN McAulay)*

Facing Page: Pilcher, Manchester's new General Manager, had barely settled in from Edinburgh when the opportunity to take the MTTA party to Paris came. Here he poses on deck, by the funnel, with AG Ellaway his MTTA Vice President seated alongside him. Behind Pilcher, and more significant than the Captain, is AW Hubble, General Manager of Crossley Motors, makers of the bus, and the one who would have had to sort out any problems with the bus! MTTA's General Secretary, J Beckett, stands between them. *(STA)*

Metropolitan-Vickers and WA Stevens of Guy Motors Limited (with reference to trolleybus experience).

The object of the Conference had been the study of, and discussion regarding, regenerative braking and also to inspect and travel on some of the ultimate 185 Type 'G' and 'L' regenerative cars of the Public Transport Company.[34] Extensive travel on these vehicles together with works visits to examine equipment were followed by M. Bacqueyrisse's paper describing the system in all its detail, and in augmenting what he had previously said at the Warsaw Convention.[35]

The initial experiments had taken place as recently as the previous year, involving rewound motors but latterly the firm of Alsthom (now Alstom) had produced a 'state-of-the-art' self-ventilated regenerative motor. These had displayed a great improvement in the elimination of excessive heat and cutting down on flashing at the commutator brushes.[36]

The original 1929 Paris conversion involved Type 'TH523' shunt-wound motors. They had been reduced to series-only working and this had, to quote M. Bacqueyrisse, as translated, resulted in 'certain inconveniences' from the traditional series-parallel state, principally in a lack of speed. This sluggishness led to further experimentation with other forms of compounding. This time the series-parallel function was used along with compounding of the motors. In the third variety, the series-parallel function was retained with the motors working in compound state while in series, but as series motors while in parallel, thereby only using the regenerative braking facility while in the series mode.[37] Bacqueyrisse used the same slipping segment idea in his controllers as Raworth had done.

The French found that using existing motors modified to compound winding fell short of the ideal because they had not been designed for regenerative braking from the start and they never quite came up to the commercial speeds required of them The 20-25% in economy in current consumption initially observed never reached more than 6-8% in maintained practice. History would repeat itself in Britain in a few years' time. Performances in Paris were never given the chance to improve there as the abandonment of the tramways was imminent. By 1935 a quarter had gone and the closure was completed on 14th August, 1938 with Versailles succumbing to the motorbus in March 1937. [38]

It was as a direct result of the Paris Conference that R Stuart Pilcher, in his capacity as President of the Municipal Tramways & Transport Association suggested that practical experimentation in regenerative braking should be carried out by the Association. He nominated a sub-committee to consider the question of collective investigation and experimentation, under himself as chairman. His team consisted of: Alderman AG Ellaway (Vice President, MTTA); AC Baker, Birmingham; JM Calder, Reading; John F Cameron, Northampton; H Gerrard, Glasgow; R Priestly, Liverpool; and C Owen Silvers, Wolverhampton. [39]

GH Fletcher of Metropolitan-Vickers, who had also attended the Conference, and had obviously been impressed, had already started researching and designing regenerative equipments himself. First of all, these were based on modifying existing series-wound motors but then with purpose-built equipment for British tramcars. The MTTA gave the first order to Metropolitan-Vickers for a set of equipment for initial experiments in Manchester.

That Glasgow should, almost immediately after the MTTA, order regenerative equipment for their own purposes, does give rise to the question – 'Why the hurry?' It would seem with the benefit of hindsight that Glasgow was indeed early in foreseeing the clamant need for a fleet of new cars of considerably higher overall standard than the modernised Standard cars, or even the so-called 'Kilmarnock Bogies' if any impression at all was to be made on the all-conquering motor-bus. The ability of such a new fleet to exploit the proven economies of regeneration over and above would have been a plumb acquisition. Here was the opportunity to test it out! It would be futile to dismiss the thought that but for the admitted failures of the later test fleet of 40 regenerative 'Standard' cars there might have been a fleet of regenerative 'Coronations!' The hope was high among tramway concerns that here indeed might be the answer for the British tramway profession in the face of the rising rout by the internal combustion engine.

THE MTTA LEADS THE WAY TO PHASE TWO REGENERATIVE BRAKING IN TRAMCARS

As has been seen, upon the return of the British delegates from the Paris Conference the President, of the MTTA, R Stuart Pilcher, set in motion the Association's purchase of a set of regenerative equipment. It is not surprising that the order for the first set went to Metropolitan-Vickers Limited. After all, their GH Fletcher had been interested enough also to attend the Conference and he had almost certainly been involved in a previous order for a regenerative braking trolleybus motor. He had already applied himself to designing a set of such machinery. As this set would – not unnaturally – be delivered to Manchester Corporation, it was a pair of their 'MV105DW' tramway motors taken from a new 'Pilcher' (or 'Pullman') car, 420 that was despatched to M-V's Attercliffe Common Works in Sheffield for re-wiring to compound, regenerative control status whereupon the motors were re-designated Type 'MV105CW' and returned for testing under Car No. 420. The set of equipment cost £500.[40] The complementary controllers, Type OK-36 B, and other ancillary equipment were built at Trafford Park. It was the realisation that the preliminary tests with this MTTA equipment in Manchester would

The French found that using existing motors modified to compound winding fell short of the ideal because they had not been designed for regenerative braking from the start

The hope was high among tramway concerns that here indeed might be the answer for the British tramway profession in the face of the rising rout by the internal combustion engine

take a long time that prompted Glasgow Corporation to place their own order for a set of equivalent equipment for themselves, as previously noted. This was virtually identical to the MTTA set and gave for reciprocity of modifications between the two, valuable to both, as testing progressed, in the early period of investigation. The Glasgow set, comprising 'MV101DR' motors was re-designated 'MV101CR' and applied to Glasgow Standard Car 305. Controllers of type OK40B were supplied to accommodate the EMB air braking interlock valve box then being fitted throughout the Glasgow fleet. Cars 420 and 305 are more fully described in the following Chapter.

THE GUY-STEVENS PATENT REGENERATIVE CONTROL SYSTEM FOR TRAMCARS

With the arrival of interest from Paris in regenerative braking in the tramway world, the trolleybus builders, who were some five or so years in advance of the tramways in this technology, made an approach to the MTTA with a view to trading their system of already advanced regenerative control. A letter was read from C Owen Silvers the Wolverhampton Manager, together with a copy of one dated 7th January, 1931 that he had received from Guy Motors Limited, recommending the purchase of a set of regenerative equipment for tramcars based on the trolleybus system that had already been so successful. Not only Wolverhampton, but also Nottingham Corporation and the South Lancashire Tramways Company, as well as several others, were using their regenerative equipment in trolleybuses. The Guy Motors letter suggested that their system of regenerative braking should be tested out in Manchester against the MTTA equipment already in use with Pilcher Car 420.

The letter reads:

you will realise that the Paris system is not the last word

> 'As you know, we are very keen on this question of regenerative control, and I venture to say that we have had more experience with this system than any other manufacturer and quite definitely to make it a success in every direction, there are some difficulties to be overcome, and the various methods of doing so are covered by our patents.
>
> Of course, you will realise that the Paris system is not the last word and in a veiled way they admit there are certain difficulties in their system which they have not overcome, which I can honestly say do not appertain to our's.
>
> In a circular letter sent to all the tramway authorities in November last, you will notice in the last paragraph we suggested that if they were desirous of trying the regenerative braking system of tramcars we should be happy to go into the matter with them and that we had some conversions in hand to a tram for one of the largest undertakings in the country – actually I might say that this is for the London United.'
>
> (Signed) Sydney S.Guy
>
> 7th January 1931

Did this eventually achieve final trial under London's tramcar 1418? Co-author opinion is to this extent, as an almost subconscious memory-trace, from early working experience!

This letter, passed through Mr Silvers – on the Committee – to the MTTA appearing in the Minutes of that Association of January 1931 as Appendix No.7, went on to mention that, depending on the type of motor to be converted, it was possible in some cases to convert the existing motors and supply new equipment at a cost of only £190. It should be remembered that the MTTA equipment had been designed around the motors of a then modern, light construction, two-motor tramcar and that as such, later proved to be very hard pressed to cope with a heavier double-bogie maximum-traction 1907 vintage London car. The trolleybus was rubber tyred, running on a far less smooth roadway than offered by the steel rail, and did require much more powerful motors. The modified, more powerful, trolleybus motors would therefore have been much more suitable for such bigger and heavier trams.

The correspondence with Guy Motors Limited, together with specifications and drawings of the Guy-Stevens patent regenerative control system put forward by Mr

Silvers, went before the special sub-committee to deal with the subject, which resolved:

'No further steps to be taken in regard to Messrs. Guy Motors Limited, and that the thanks of the Council be tendered to them for the trouble they had taken in the matter'.[41]

This failure to assess the trolleybus regenerative system of braking on tramway cars could have been a missed opportunity for both the tramways and for the Guy-Stevens regenerative motor which obviously had a developmental lead of at least five years over the Fletcher-Metro-Vic machine that was then only a largely experimental motor. The trolleybus version had had the opportunity already to face, and iron out, a lot of elementary troubles! Again with hindsight and in no way to excuse the rather miserable decision of the MTTA, no matter how advanced the Guy-Stevens regenerative equipment might have been, inevitably regeneration had to bide time for the maturity of solid-state electronics as an integral factor in the mastery of the system, to enable it to attain its acme of perfection and useful function. One can only speculate as to what the future of the Guy motors might have been if their regenerative braking had found a widespread following. Today, they are probably best remembered as builders of utility buses during World War II.

Anyhow, and whatever way one looks at it, there is no doubt that the MTTA's action was thought to have been in the best interests of the tramway movement in the face of rising opposition from the internal combustion vehicle for it did lead to a string of quite

Nottingham Corporation turned to Park Royal and Karrier for their 1931 trolleybuses intake. General Manager WG Marks was not an advocate of regeneration as later became apparent when he moved to Liverpool. *(STA)*

Following the success of the 'Diddler's' conversion, London opted to specify regenerative braking for its large standardised pre-war fleet. Regenerative control was more successful when designed as part of the original specification rather than retrospectively fitted. *(STA)*

Crossley supplied TDD4 (four-wheeled) and TDD6 (six-wheeled) trolleybuses to their neighbouring corporations of Ashton and Manchester in the late 1930s. RS Pilcher, following his experience with Manchester tramcar 420, and a desire to force down traction costs, specified regenerative control. Crossley came late into the market, no doubt determined that Guy and Sunbeam, or Karrier, or any one else, were not going to get a look in!

(Both courtesy GMTS Museum, Manchester)

intensive individual system investigations into regenerative braking. All of these, while ultimately reverting to a simple series-wound working for various reasons, and over shorter or longer periods of activity, have been treated in the following chapter under the heading 'The Second Phase of Regenerative Braking in Tramcars'.

To close this chapter it is not out of place to provide a thumb-nail sketch of the well-known Guy Motors Limited firm of builders. They had designed and had converted for London United, one of the largest undertakings in the country, at least two compound-wound regenerative motors for tramway purposes.[42]

Sidney Slater Guy founded Guy Motors Limited, of Wolverhampton in 1914,[43] having previously been Works Manager of the Sunbeam Motor Company Limited, also of Wolverhampton.[44] The mid-1920s saw Guy Motors building motorbuses in Wolverhampton and were the very first to produce a diesel-driven bus, in 1929.[45] Diversification into trolleybus manufacture, with regenerative control, in 1926,[46] signified the firm's association with WA Stevens, of that interest, producing the first six-wheel top-covered double-decked trolleybus with open rear staircase and regenerative braking control – the well known No. 33 in the Wolverhampton fleet.[47] In 1948 Guy Motors Limited bought the Sunbeam Trolleybus manufacture interests which, commencing in 1931, had been taken over by the Rootes Group in 1935.[48] Sunbeam had the doubtful distinction of being the supplier of the last production batches of trolleybuses in the country with a final batch for Bournemouth Corporation in 1962. It fell, however, to the pioneering Bradford Corporation system to operate the last UK trolleybuses in 1972.

Facing page: reproduction of an advertisement from *Passenger Transport* magazine for March 1938. *(STA courtesy DS Hellewell)*

REFERENCES: CHAPTER 4

1. Westbrook, M.H.: The Electric Car: The Institution of Electrical Engineers, London: 2001: Ch-2:p18
2. Kennedy, R.: Electrical Installations: 1909 (London): p134
3. Tramway & Railway World: 1905 – August 10: p135: Col.2
4. The Tramways Act, 1870, 33 & 34 Vict., Ch78: Clause 19
5. Tramway & Railway World: 1902 – August 7: p286
6. Ibid: 1905 – August 10: pp135-139
7. Kennedy, R.: Electrical Installations: 1909 (London) pp134-137
8. Personal Correspondence: I.A.Souter to J.D.Markham, 1-6-2003
9. Ibid: T&RWld: 1905 – August 10: pp135-139
10. Ibid
11. Yearsley, I.: Tramway Review: 1974 – Spring:10:77:131
12. Rush, R.W.: Tramways of Accrington 1886-1932: 1961 - L.R.T.L.: pp27-28
13. Ibid: p79
14. Markham, J.D.: personal communication
15. Pott, M.A.H.: Valeur respective des Tramways, Autobus et Omnibus a Trolley, comme moyens de Transport: In Revue Manselle: 1813 –Aout: No.80: L'Industrie de Tramways et Chemins de Fer (Organe de L'Union des Tramways et Chemins de Fer d'Interet Locale de France: p325.
16. The Railless Electric Traction Company Limited, London: Trade Introductory pamphlet reprinted from the Tramway & Railway World
17. Ibid
18. Hal, C.C.: Rotherham & District Transport: Vol.3: 1939-1947: Rotherham -1999: p76
19. Pilcher, R.S.: Presidential Address at M.T.T.A. Conference, Southampton, 24-26 June, 1931 – on the Regenerative System
20. Groves, F.P.: Personal communication 24-9-2002
21. A Pictorial History of Halifax Transport: undated: unacknowledged, but well illustrated: A4 size pamphlet: possibly issued by the Local Authority
22. The Guy Motors Ltd., Trolleybus Promotional booklet: 1930 – March: p4
23. Bishop, R.A.: The Electric Trolleybus: No date: (Pitman Press No.C1-(5687): Chapter III: p46
24. Bett, W.H. & Gillham, J.C. (Editor R.J.S.Wiseman): The Tramways of the West Midlands (Network Edition): L.R.T.A.: No date – but 1999 quoted in text: pp8-9
25. The Bradford Trolleybus Scheme: in Tramway & Railway World: 1930 – 19 July: LXVIII: p45
26. Ibid: Guy Trolleybus Promotional Booklet: p5
27. Parker, T.C. & Robbins, M.: A History of London's Transport: Vol. II – The Twentieth Century to 1970: 1974: Chapter 12: p241
28. Blacker K.: The London Trolleybus – Vol. 1 – 1931-45 Capital Transport 2002: Chapter 3: p33
29. Groves, F.P.: Personal correspondence: 15 June, 2001
30. Ibid: Groves, F.P.: 26 September, 2002
31. Tramway & Railway World: 1930 – November 13: p283: Col.1: Mr W.A.Stevens on Regenerative Control with Guy Trolleybuses.
32. Ibid: p281
33. Ibid: pp303-4
34. Robert, J.: Les Tramways Parisiens: 1959: Paris:2Edn.: Chapter – "Le Materiel Roulant de la S.T.C.R.P." (Societe de Transports en Commun de la Region Parisienne): pp206-218
35. Tramway & Railway World: 1930 – November 13: pp281-282
36. Ibid: Robert, J: Les Tramways Parisiens: p218
37. Ibid: 218
38. Ibid: Unnumbered page: "Suppression des Lignes de Tramways par Ordre Chronologique"
39. Tramway & Railway World: 1930 – December 18: p342
40. Markham, J.D.: personal communication: 28-8-2002
41. Appendix No.7, of 7th January, 1931, to the Minutes of the Municipal Tramways & Transport Association, being a letter from Mr Sidney S.Guy to Mr C Owen Silvers – Memb. M.T.T.A. Committee – for presentation to the M.T.T.A. at their next meeting.
42. Hannay, R.: Guy Motors and the Wulfrunian: 1978: Transport Publishing Company: foreword
43. Pettie, J.: Guy Buses in Camera: 1979: Ian Allan Ltd.
44. Ibid: Hannay, R
45. Ibid: Pettie: The Early Years: p10
46. Groves, F.P.: personal communication
47. Ibid: Pettie, J.: Guy Buses in Camera: Chapter – Post War Developments: p44
48. Ibid: Pettie, J.: p44

ELEVEN YEARS SERVICE

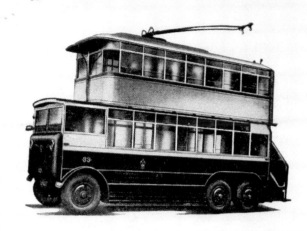

The first 6-wheel Double-Deck GUY Trolley Bus (shown on the right) equipped with regenerative control, has been given an honourable retirement after eleven years service for the Wolverhampton Corporation, who now operate a fleet of 99 GUY Trolley Buses.

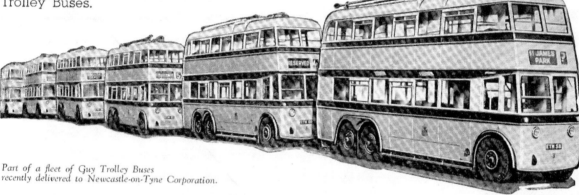

Part of a fleet of Guy Trolley Buses recently delivered to Newcastle-on-Tyne Corporation.

REGENERATIVE CONTROL

In the GUY system the motor automatically becomes a generator, sets up a counter-electromotive force and returns current to the overhead wire, this providing a valuable retarding effect on down gradients, reducing wear of brakes and economising current.

Some of the tests record a saving of 30% in current consumption as compared with non-regenerative systems. It is not claimed for the GUY regenerative system that it effects this saving in every service, but it is claimed that running under exactly the same conditions as a machine fitted with series motor, it is possible to effect a saving in current consumption of 15% to 30%, depending on the local conditions.

The motor is capable of providing smooth and rapid acceleration; when regenerating it provides progressive retardation at all road speeds down to 12 miles per hour.

THE SECOND PHASE OF REGENERATIVE BRAKING IN TRAMCARS

As has been seen in the previous Chapter, despite the 1911 embargo imposed by the Board of Trade on regenerative braking in trams, the concept was developed in trolleybuses to the benefit of tramways twenty years later. Struan Robertson explains how the Municipal Tramways and Transport Association ultimately took a grip of the concept once more and led the way to the various experiments in the 1930s.

Tramway systems had suffered heavily under the use of regenerative braking during the elementary, or First Phase, of its use. The concept had been researched by many at the turn of the 19th/20th century, but the work of John Smith Raworth had made the greatest impact on a technique that was well ahead of its time. Raworth is much to be admired for rendering his surmise practical. However, as recorded, several run-away accidents of greater or lesser severity, coupled with a far higher than average incidence of breakdowns requiring recovery, involved regenerative tramcars on the Birmingham & Midland Company when running on Birmingham Corporation tracks in 1905-06. This suggested that faults of the shunt coils or their connections had been a regular and unwelcome feature. Due to this unacceptability, Birmingham Corporation forbade the use of these cars within the city as from 1st January, 1907.[1] As previously seen, Halifax had an almighty runaway collision between its two Raworth demi-cars. (Appendix 1 gives more full information on this accident). The overhead current had been cut off while one of them was descending a hill under its regenerative brakes thus breaking the regenerative circuit completely. Ultimately, as explained, it was the 1911 Rawtenstall runaway accident that led to the Board of Trade recommending the withdrawal of this system of braking, entirely.

It was against this background that further attention to the development of regenerative braking, in tramway usage, was dropped. This was followed by the Great War and its enforced cessation of experimentation. Then there was the post-war depression with its almost total lack of materials and funding. Even despite the 1926 successes with regenerative braking in trolleybuses, tramway concerns were simply not interested. In any case, the Board of Trade had not lifted its sanctions against regenerative braking.

The Paris experience, coupled then with the covert awareness, already, of the successes in trolleybus usage, was an immense stimulus to the tramway trade, suffering as it was, from escalating competition from internal combustion vehicles to the point of boded extermination of the tramway movement altogether. Could this system perhaps be the saviour of British tramways? Sadly, even with all the advances of electric technology at that time, it turned out not to be. However, the force of this stimulus and the need to do something about the soaring motor trade opposition crashed through all respect of Board of Trade sanctions relating to regenerative braking. The result was that some of the bigger tramway concerns forced on regardless with experimentation with regenerative braking, fully backed by the MTTA. It should be remembered here that the

(the) . . . political climate was generally hostile to trams now,

political climate was generally hostile to trams now, with the Royal Commission's 1930 Report coming out against the tram and recommending that they should be replaced by buses (or trolleybuses?).

Manchester led the way, followed by London and Glasgow, neck and neck, then Edinburgh, Leeds, Halifax and, finally and largely by chance, Birmingham. Liverpool ordered a set of equipment but withdrew in the course of its installation. The Brighton Corporation Tramways, under William Marsh, had agreed to the trials of equipment that two firms in the south of England had recently commenced building.[2] This latter venture never matured due to their suspending development of regenerative control for the time being.[3] No further detail relating to these firms has become available. Although never named, they were almost certainly Messrs Crompton Parkinson of Chelmsford and Messrs Allen West of Brighton. It is possible that the Brighton equipment was a duplicate of that almost commissioned on Liverpool 810.

MANCHESTER 'PILCHER' CAR 420

As was seen in the previous Chapter, the MTTA decided to purchase one set of regenerative braking equipment from the Metropolitan-Vickers Electrical Company Limited at a cost of £500. This was for the purposes of experimentation for the information of the Association. It was decided that these experiments should be conducted in Manchester under the supervision of RS Pilcher, President of the Association. It was also open to members to borrow the equipment for trials on their own metals.

Car 420 was then only four months old, and to the design of Mr Pilcher. It consisted of a lightweight double-decked car on a four-wheel truck with MV105DW

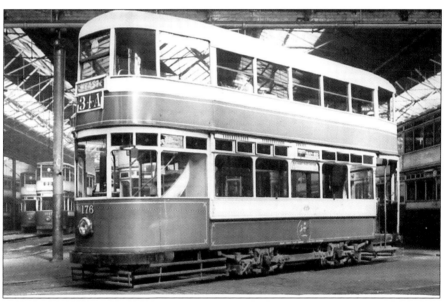

Car 176, a Manchester "Pullman" or "Pilcher" car. This was of the same class, and identical, apart from the traction equipment, to the Manchester regenerative braking car 420. Unfortunately, the search for photographs of Car 420 in Manchester has proved fruitless.
(Photo: STA)

Thanks to Mr Ian Souter, and through him, Mr David Heathcote whose late father took this photograph, it is possible to illustrate Manchester 420 in the later guise as Aberdeen 41. The car ran there from 1947 until October 1955 and is seen at the Bridge of Dee terminus around 1948. Some of the alterations carried out after arrival in Aberdeen can be seen.

motors re-designated 'MV105CW' on being rewound for compound working on the regenerative braking phase and was selected as the Test Car motor. The standard car against which the tests were performed was 380, a sister car of the same design. Like the other members of the class it had two MV105DW simple, series-wound motors each of 50 hp on a one-hour rating. The two cars were paired together from 21st September until 4th October 1931, both dates inclusive, for initial assessment. They ran on the Belle Vue and Weaste route. As Weaste is in Salford it is implicit that the Salford Corporation officials consented to the use of 420 on their system. Manchester Corporation invited General Managers and/or their Engineers, both of tramways and of trolleybus operators, to visit Manchester and inspect the Regenerative Control at work. For this all necessary facilities were offered to them.[4]

The equipment had been in regular public service throughout the year and reports were issued from time to time to the Members of the Committee dealing with the experience gained from operation over some 25,000 miles. The first interim report covered about 4,400 miles and a saving in energy consumption over the standard car of 14.9% was achieved. This was mostly with drivers under training in the use of the new equipment. In the second report the saving had increased to 24%, the mileage being about 10,700. The third report summarised the results for some 9,900 miles of operation and showed a saving of 16.3%. For the entire mileage of 25,000 the overall saving was 19.4% and the experiments continued apace. In addition to these reports, the Committee had been informed of the experience of the Glasgow Corporation Transport Department with similar equipment supplied by the same manufacturers.[5]

The next report was issued after 31,000 miles and indicated that Manchester had covered eight different routes, clearly with different characteristics, and always in comparison with a similar car with standard equipment. For this mileage the overall saving in energy consumption amounted to 19.8% of that of the standard comparison car. With Glasgow's similar equipment the saving in energy exceeded 30% while Edinburgh's had shown a greater energy economy even than that.[6]

Manchester 420 had the equipment taken out in 1932 and it was passed to the London County Council Tramways for further assay. It was to return to Manchester later.

Metropolitan-Vickers Series-Parallel Regenerative Tramway Schematic Diagram for the Municipal Tramways & Transport Association's Experimental Regenerative Tramway Scheme.[7]

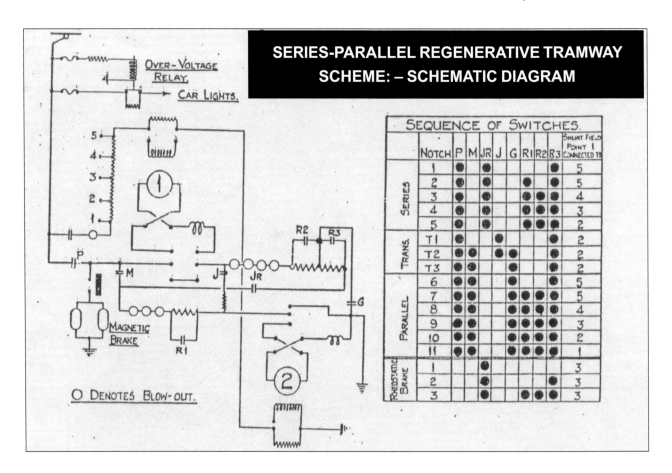

SERIES-PARALLEL REGENERATIVE TRAMWAY SCHEME: – SCHEMATIC DIAGRAM

OVER-VOLTAGE RELAY.
CAR LIGHTS.
MAGNETIC BRAKE
O DENOTES BLOW-OUT.

	NOTCH	P	M	JR	J	G	R1	R2	R3	Shunt Field Point 1 Connected to
SERIES	1	●	●	●					●	5
	2	●	●					●		5
	3	●	●				●	●	●	4
	4	●	●				●	●	●	3
	5	●	●				●	●	●	2
TRANS.	T1	●			●				●	2
	T2	●	●		●	●			●	2
	T3	●	●		●				●	2
PARALLEL	6	●	●			●			●	5
	7	●	●			●	●		●	5
	8	●	●			●	●	●	●	4
	9	●	●			●	●	●	●	3
	10	●	●			●	●	●	●	2
	11	●	●			●	●	●	●	1
RHEOSTATIC BRAKE	1			●						3
	2			●				●		3
	3			●			●	●	●	3

SEQUENCE OF SWITCHES.

GF Sinclair's Braking and Acceleration Tests involving Manchester Regenerative Car 420

While 420 was still working on home ground, GF Sinclair, Rolling Stock Engineer of the London County Council Tramways – courtesy of Manchester Corporation Tramways – drew up a series of comparative figures on acceleration and braking tests at various speeds with a trio of Manchester cars as set against an equal number of LCC cars. Of the former, one was the MTTA regenerative braking car 420.

The individual cars participating were:

'A' LCC 110 Class HR/2 4-motor (2 in permanent series) with 4 magnetic brake magnets and wheel block attachment.
Weight 17 tons 7 cwt

'B' LCC 995 Class E/1 2-motor maximum traction bogie with 4 magnetic shoes and wheel block attachment.
Weight 16 tons 8 cwt

'C' Manchester 380 'Pilcher' 2-motor single truck type with 2 magnetic brake shoes and no wheel block attachment.
Weight 11 tons 1 cwt

'D' Manchester 420 'Pilcher' same as Car No.380 but with Regenerative Braking.
Weight 11 tons 1 cwt

'E' Manchester 421 a 1926 2-motor maximum traction bogie fitted with air braking and track shoes 17in long.
Weight 14 tons

'F' LCC 1538 Class E/1 similar type and weight to Car 995 but arranged for rheostatic braking on the first 3 notches and magnetic braking (without the wheel block attachment) on the last 4 notches.
Weight 16 tons 8cwt.

All the tests were made with good rail conditions on practically level track. Anywhere that a slight gradient occurred, tests were made in both directions to correct the grade error. As far as possible the controller operation was the same, the aim being to obtain the maximum possible acceleration and retardation without wheel-slip – no sand being used throughout the tests. The figures given for average accelerations from rest to each of the speeds are in feet per second per second.

Speed mph	A Car 110	B Car 995	C Car 380	D Car 420
5.0	4.0	3.45	3.2	3.2
7.5	3.9	3.4	3.2	3.2
10.0	3.8	3.25	3.0	3.0
12.5	3.6	3.0	2.75	2.75
15.0	3.4	2.7	2.5	2.45
17.5	3.2	2.45	2.25	2.15
20.0	2.8	2.2	2.0	1.9
22.5	2.4	1.9	1.8	1.7
25.0	2.0	1.75	1.6	1.5

Note that, as would be expected, the acceleration of the car with regenerative equipment was virtually the same as with the similar car with series-wound motors. These speeds were the averages of at least three tests for each speed from 5 – 25 mph. There was an increased current consumption in service with 4-motor equipment than with 2-motor equipment operating under the same conditions and approximately the same car weight. Taken over a period, the average consumption was 2.6 units per car mile for the 4-motor car as compared to 2.4 units for the 2-motor car. Calculated as a percentage this increase was small compared with the percentage increase in acceleration. The principle of driving on all four axles was conceded to be the most efficient and it was the capital cost of the equipment that was the deciding factor. With efficient low-cost 2-motor equipment driving on to all four axles, an alternative could be found. It was worthy of note that there were already several experiments being made to develop such a drive on as simple lines as possible. In many ways the conclusion matched that of Henry Mosley, General Manager of the Burnley Tramways, round about 1906.

The need for very high retardation brakes cannot be overestimated, especially as in the 1930s car speeds were being allowed to increase to 30 mph. It was generally accepted that the average coefficient of static friction for all weathers between wheel and rail surface was 0.16, and that the retardation of the car when wheel, or disc brakes, only, were used depended upon the speed of application of the wheel brake and the facilities for adjusting the pressure. Air or mechanical track brakes fitted with rolled steel shoes having an average coefficient of 0.2 were often used. In both types of braking the available adhesion defines the limit of the decelerating force. It also applies to rheostatic braking and regenerative braking – all having a definite limit of retardation depending on the coefficient of static friction between the wheel tread and the rail.

With the magnetic track brake, an additional retarding force was obtainable from the drag of the shoes on the rail independent of the weight of the car. To retain as closely as possible the high retardation of magnetic braking throughout and also relieve the track of some of such a destructive form of braking, must be the main objective. With rheostatic braking the peaks that occur when cutting out resistances always cause a tendency to skid the wheels at high car speeds. With magnetic braking the same tendency to pick up the car wheels does not exist for the same overall rate of retardation. It was deemed necessary, then, that the peaks of rheostatic braking could be dampened down and thus overcome the then disadvantage and extend the practical use of rheostatic braking. The more energy absorbed by the motors and resistors the less destruction of the track. This was quite the most costly element of a tramway system!

AVERAGE RETARDATION IN FEET/SEC/SEC

Speed mph	A Car 110	B Car 995	F Car 1538	E Car 421	D Car 420	C Car 380	G Car 420
5.0	6.4	5.3	5.3	4.6	4.6	3.25	3.25
7.5	6.8	5.5	5.5	4.9	4.9	3.5	3.5
10.0	7.0	5.6	5.6	4.9	4.9	3.7	3.7
12.5	7.0	5.7	5.7	4.8	4.8	3.75	3.7
15.0	6.9	5.8	5.7	4.8	4.6	3.8	3.65
17.5	6.6	5.7	5.5	4.6	4.5	3.75	3.6
20.0	6.2	5.6	5.3	4.6	4.33	3.75	3.55
22.5	5.8	5.4	5.1	4.5	4.2	3.75	3.5
25.0	5.7	5.3	5.0	4.4	4.0	3.6	3.4
	LCC	LCC	LCC	Manch	Manch	Manch	Manch

STOPPING DISTANCE IN FEET

Speed mph	A Car 110	B Car 995	F Car 1538	E Car 421	D Car 420	C Car 380	G Car 420
5.0	4.2	4.5	4.5	5.9	5.9	8.5	8.5
7.5	9.0	11.0	11.0	12.0	12.0	17.5	17.5
10.0	15.5	19.5	19.5	22.0	22.0	30.0	30.0
12.5	23.0	29.0	29.0	31.0	35.0	45.0	45.0
15.0	35.0	42.0	42.0	50.0	52.0	65.0	66.0
17.5	49.0	58.0	60.0	70.0	71.0	89.0	91.0
20.0	69.0	78.0	81.0	92.0	100.0	117.0	120.0
22.5	93.0	100.0	101.0	120.0	127.0	148.0	156.0
25.0	120.0	125.0	133.0	150.0	165.0	181.0	196.0

Regarding the regenerative braking equipment of Manchester Car 420, each power notch had a definite running speed dependent upon the motor field excitation. That was an advantage in service. The checking of the car speed required no actual "brake" application as the motors regenerated to the overhead until the desired speed was reached. The loss of braking force when passing over the transition and "off" positions was a real disadvantage. The retardation obtained in regenerative braking, only, was

The checking of the car speed required no actual "brake" application

78

equal to the rheostatic braking but when the magnetic braking was brought in on two notches, the retardation of the regenerative equipment improved and approximated to that obtained with the car fitted with the air-brake equipment.

THE LONDON (LCC) CLASS E/1 CAR NUMBER 779 — 1932

In October 1932, at the request of TE Thomas, General Manager of the then London County Council Tramways, the Association's two experimental regenerative motors were transferred from Manchester to London for trials on the London conduit system. The equipment had originally been designed for conditions less severe than those of the London County Council system. As a result, the performance in terms of saving energy was not as good as previously obtained in Manchester. Nevertheless it yielded many useful and interesting results.

The London County Council bogie cars with which Car 779 was compared from October 1932 until May 1933 mainly had 60-63 hp motors although some, like 901 (a north-side car) still retained their original Westinghouse No.220 motors that had a one-hour rating of 42 hp.[8]

The main features of Mr Thomas' report were directed to matters other than the energy saving, but they were of equal importance on a high-speed tramway system, and indicated that:

1. Compound motors of robust design, capable of regenerating down to a speed of around 6 mph should be as reliable in tramway service as series motors.
2. The use of this type of motor on the conduit system, where frequent reversals of polarity were encountered, offered no serious handicap to traffic, or damage to the motors.
3. The regenerative system offered the advantages of:
   a. Increased schedule speeds with lower energy consumption
   b. Considerable relief in the wear of track and the magnetic shoes
   c. Assistance to traffic on account of the larger number of running speeds and a better electrical control than with the series-parallel system.
4. The acceleration of an LCC car, fitted with the experimental motors compared very favourably with that of the standard LCC car.
5. When used with the magnetic brake the retardation obtained with the experimental motors was quite equal to that of the standard LCC car.

London County Council (L.C.C.) Tramways Type E/1 car 901 here represents its sister car 779 – London's first regenerative braking car. Both were identical in external appearance.

(Courtesy: National Tramway Museum Archive)

The London E/1 Class maximum traction Car 779 underwent gruelling night-time tests and daytime service. It had done very well despite the drawback of a near 50%

increase in unladen car weight that had not been allowed for in the original equipment rating provided. Magnetic brake applications were reduced by almost 50% and the car's magnetic shoes lasted almost 4½ times those of a standard vehicle. These factors determined the LCC engineers to stage a formal investigation into a service of modern custom-built regenerative motors by different makers.[9]

As a result of these favourable results with the Association's (underpowered) set, the (by then) London Passenger Transport Board (LPTB) ordered regenerative equipments specially designed for their local conditions and made further tests. These sets included motors of 60 hp one-hour rating at 525 Volts providing a lowest balancing speed of 5 mph. They were probably the largest compound motors yet built in this country for rail traction. Service tests showed an energy saving of 22 % over a period of three weeks, when compared with normal equipment.[10]

Mr Fletcher was an extremely busy man in those days, travelling between Manchester, Glasgow, Edinburgh and London grooming his M-V regenerative equipments in these cities and nursing them along, then back to his office in Sheffield. A series of correspondence from Metro-Vick, at that time has come down, illustrative of the problems inherent with the regenerative equipments of this Second Phase of their application.[13] The much advanced technique of the 1930s was, even then, short of adequacy in extracting the full benefit of the regenerative process. This had to wait until the advent of electronic control for its maturity.

The first letter, signed by RP Knight, Traction Engineering Department of the Metropolitan-Vickers firm, applied to Manchester 420 shortly before the transfer of the MTTA equipment from Manchester to the LCC. This highlights a lot of difficulty with the shunt field notches, with overheating. The letter of 9th May, 1933, when the equipment was at work in London, shows that further modification of the controller had been necessary. These were aimed at increasing the rheostatic brake power following cessation of regeneration together with adapting it for London's conduit system. This was by rearrangement of the shunt field connections to convert the shunt fields to parallel, rather than in series, and a further modification of the values of the main resistor.

Magnetic brake applications were reduced by almost 50% and the car's magnetic shoes lasted almost 4½ times those of a standard vehicle

METROPOLITAN-VICKERS ELECTRICAL COMPANY LIMITED,
ATTERCLIFFE COMMON WORKS, SHEFFIELD.

FROM .. Traction Motor. Engg.. Dept.   TO   Chief Engineer,
                                            Traction Control   Dept.

SUBJECT Manchester Regenerative Tramcar. Attention of .. Mr. Brooks.
                                            (Mr. Ramsden)

RPK/G.

7th October 1932.

With reference to your letter of the 5th inst and our subsequent telephone conversation, we enclose herewith two curves (Proposals A & B) showing the suggested arrangement of the shunt field notches on the modified Manchester car.

Proposal A.

This is the same as the original equipment (Curve No. 535526 D) except that the shunt on Notch 5 has been reduced to 1.3 Amps and the 1 amp notch in parallel replaced by 2 notches with 1.3 and .7 shunt amps.

For this proposal we suggest that the excitation on the resistance notches should be the same as on the existing equipment, i.e. 6.35 amps in series and 4.2 amps in parallel.

Proposal B.

Here the whole of the notches have been rearranged to give rather more regeneration.

With this grouping it will probably be advisable to reduce the excitation on the parallel resistance step to 3.4 amps to avoid possible trouble in backward transition, but the series resistance steps could be arranged for 6.35 amps excitation if found convenient.

We think that the choice between the two proposals should depend to some extent on the service in which the car is ultimately to operate. If it is still to be used in Manchester where there are no steep gradients and where severe retardation is not required in normal service, proposal A would probably be more satisfactory, particularly as it is known that it gives satisfaction.

The "Mr Sheers" referred to in the above correspondence was a specialist traction motor design engineer at Metropolitan-Vickers, Attercliffe.[11]

Manchester Regenerative Tramcar.   7th October 1932.

We would suggest designing the shunt control resistance for a nominal shunt field resistance of 37 ohms per motor which, on the 6.35 amp notch would give an extreme range of 8 to 5.4 amps at 500 V.

Mr. Sheers will be in Manchester tomorrow, Saturday, and would be pleased to discuss with you any of the above points..

TRACTION MOTOR ENGINEERING DEPT.

Manchester Regenerative Tramcar.   7th Oct. 1932

Proposal B would be more suitable for steep gradients or for fast schedules, but it leaves less margin of safety against flashovers in emergency conditions. The commutation of the Manchester Motors is not as good as on the Glasgow and Edinburgh machines and although we consider that they should meet the conditions of Proposal B, we should hesitate to recommend its adoption unnecessarily.

We are preparing the information you require regarding rheostatic braking and we hope to post it to you tomorrow.

MOTOR RESISTANCE.

The following table gives particulars of the motor resistances at various temperatures.

Machine No.	14,896	14,942.
At 0°C.		
Armature and I.P.	.4510	.4440
Series Field.	.0484	.0483
Shunt field.	29.1	29.4
At 20°C		
Armature and Interpole	.4900	.4820
Series Field.	.0526	.0525
Shunt Field.	31.6	31.8
At end of 1 Hour Test Run.		
Armature and I.P.	.6950	.6830
Series Field.	.0655	.0655
Shunt Field.(60°C)	36.5	36.8

In normal service the shunt field temperature will probably be somewhat higher (say 70°C) and the resistance would then be about 38 ohms, while under abnormal conditions, the shunt field resistance might rise to about 44 ohms per motor (at 120°C)

M.T.A. REGENERATIVE EQUIPMENT.

REPORT OF CHANGES TO CONTROL SYSTEM FOR

LONDON SERVICE.

provide:- The controllers have been modified so as to

(a) Six rheostatic brake notches with the motor
armatures in parallel, instead of three
with the armatures in series.

(b) New method of transition between the series
and the parallel connections.

(c) Remote control of six small contactors to
regulate the shunt excitation of the motors.

(d) Addition of three extra fingers and contacts
on the power-brake drum to isolate the negative
feed as well as the positive, at the off and
the rheostatic brake notches so as to make the
equipment suitable for the conduit system.

Drawing B.588605/4 shows all these alterations.

The shunt field connections have been re-arranged
so that the fields are connected in parallel instead of in
series, and the discharge resistance now does duty as a
series resistance to give the same shunt current values
as before, now switched off when the controller main handle
is brought to the "off" position.

The main resistance is now in two parts insulated
from each other, R1 - R4 being the starting resistance and
JR - TR being a resistance which is brought into circuit
only during transition.

The values are:-

R1	R2	1.0	ohms.
R2	R3	1.2	ohms.
R3	R4	1.6	"
JR	TR	4.5	"

WBGC/BMV.

[Courtesy: F. Philip Groves]

Note 'WBGC' are the initials of W. B. G. Collis, a
design engineer with Metro-Vic's controls department
of Trafford Park while "OK36B" was probably the
type of controller used for the London trial.[12]

This page: Comparative Results obtained by Manchester Corporation Tramways.

(Reproduced from The Tramway and Railway World: 17th September, 1931: page 164)

COMPARATIVE RESULTS OBTAINED BY MANCHESTER CORPORATION TRAMWAYS ON REGENERATIVE AND STANDARD CARS FITTED WITH MV.105, CW AND DW MOTORS RESPECTIVELY

Grade approx. 1 : 500.   Results averaged for up and down grade.

Length of Run.	Speed M.P.H.		Regenerative Car.						Standard Car.			
	Schedule 5 sec. Stop.	Maximum.	Units per Car Mile.			% Regeneration.	Acceleration. M.P.H.P.S.	Retardation. M.P.H.P.S.	Units per Car Mile.	Acceleration. M.P.H.P.S.	Retardation. M.P.H.P.S.	Percentage Saving.
			Motor.	Regeneration.	Total.							
Yards. 150	9·67	22	3·21	1·28*	1·93	40*	2·08	1·98	3·08	2·23	1·79	37*
,,	9·58	19·1	2·37	0·71	1·66	30	2·25	2·00	2·6	2·4	1·80	36
,,	8·87	16	1·91	0·45	1·46	22	2·25	1·92	1·91	2·5	1·67	23
,,	7·01	9·8	1·15	0·11	1·04	9	2·30	1·55	1·13	2·88	1·40	8
220	10·9	24·5	2·58	1·11	1·47	43	1·98	1·48	2·42	2·63	1·35	39
,,	10·7	19·8	1·70	0·65	1·06	38	1·93	1·40	2·11	2·25	1·3	43
,,	10·1	15·9	1·38	0·38*	1·00	28*	2·0	1·55	1·5	2·45	1·42	33*
,,	7·6	10·0	0·88	0·09	0·79	10	2·08	1·60	1·03	2·32	1·35	23
300	14·2	25·9	2·25	0·70*	1·55	31*	2·30	1·85	2·08	2·35	2·18	26*
,,	12·6	19·2	1·62	0·48	1·14	30	2·07	2·03	1·41	2·35	1·55	19
,,	11·4	15·9	1·23	0·16*	1·07	13*	2·25	2·00	1·09	2·45	1·70	2*
,,	8·1	9·9	0·80	0·06	0·74	8	1·90	1·50	0·71	2·90	1·25	−4
250 metres (273·5 yds.)	13·5	25·0	2·41	0·99	1·42	41	2·15	2·00	1·98	2·6	1·8	28

* Test curves show regenerative controller to have been returned to "off" position at speeds of the order of 13 M.P.H.   The percentage regeneration is therefore low, since the car energy from 13 M.P.H. to the minimum regeneration speed is lost in the hand brakes instead of being returned to the line.

[The Tramway and Railway World: 1931, September 17th: page 164].

THE GLASGOW INITIAL REGENERATIVE EXPERIMENT – CAR 305 – 1932

Returning from Paris towards the end of October 1930 from the meeting of the International Tramways & Light Railways Union were, amongst some 35 Members, two from the Glasgow Corporation Transport Department. They were Lachlan MacKinnon, the recently-appointed General Manager, and his senior Electrical Engineer, H. Gerrard. They had just had a most fascinating experience of wide-open professional welcome in Paris. They had listened to the most erudite paper by M. Bacqueyrisse, the Director of the Paris Tramways, on the Paris regenerative tramcar fleet. They had participated in open discussion on all the pros and cons and had driven around for test purposes on a fleet of those cars put at their disposal. They had visited the Paris Car Works to inspect and investigate the workings of these cars and their compound-wound machinery. They had been thrilled at the potential offered by regenerative braking, the kindness and open-heartedness of welcome they had all received and were, not unnaturally hooked on having a go at the regenerative braking process themselves (as were quite a lot of others).

As previously described, the MTTA had bought a set of 'Second Phase' regenerative braking equipment.

Glasgow, however, had ordered a set for itself. As with Manchester, a pair of MV101DR motors was sent to Sheffield for re-winding to compound fields, virtually identical to the Manchester set. Re-categorised as MV101CR, they were returned to Coplawhill Works and fitted to the next available car in for major overhaul – 305. Over a period of four months the Glasgow regenerative car covered 17,600 miles and showed an energy saving of 32% as compared with two Standard cars, each of which operated over approximately the same distance. Savings on individual days were never less than 20% and reached a highest value of 44.5%.[14]

The main details of the conversion of this box-type, self (fan) ventilated 600 Volt traction motor were:- a 25 slot armature taking coils of 3 turns per coil, including a 149 bar commutator and compound-wound fields made up of 4 main poles with 15 series turns and 3 interpoles with 66.5 series turns. The total number of shunt turns on the main poles was 760. The armature was mounted on Hoffmann roller bearings at both ends and weighed 375 lbs. The total weight of the motor, without gears and gear-casings, amounted to 1,590 lbs. With Class 'B' insulation the bench tests showed 36 hp on a one-hour rating at 120°C, working at 500 Volts, and 62 Amps, at a speed of 560 rpm and showing an efficiency of 86%. The shunt amperage was 0.5 Amps.[15]

The *Transport World* of 18th January, 1934, pp37-38 described this car as 'operating in Glasgow during the whole of 1933 with the new equipment' implying that conversion to regenerative braking control must have been during the latter half of 1932.

Car 305 was returned to service in January 1933 and ran for around 10 months as a manually controlled regenerative braking car. In November, electro-pneumatic control was installed for a further 10 months after which manual controllers were re-installed – possibly a new set of OK-40 B type. The car was stored between February and May of 1935 after which conventional controls and braking were fitted. It was eventually scrapped in 1954.

At the re-establishment of manual regenerative braking control following withdrawal of the electro-pneumatic control

A) It is possible the original, regen-converted controllers were re-used. They would have been retained at Coplawhill during the period of 10 months

B) This suggestion is less likely, being reasoned surmise, involving the possible fitting of MV-modified BT-H.B510 controllers, photographs of which had been seen at Trafford Park by JD Markham, who believes that they were associated with trials in Edinburgh.

C) Type MV OK-40 B controllers are much more of a probability because
  · They were the newest at that time
  · They were purpose-built

Two, separate, Coplawhill Works Record Books have entries recording their being fitted. However, a March 1935 Drawing – MV B589494 – could have been drawn 'after the event' in order to record further modifications.[16]

While not an outstanding success, 305 was far from being a failure, garbed as it was in quite a plethora of excessive, untried, electrical finery, for the Tramways Committee was sufficiently impressed by the overt promise of economy in many

Over a period of four months the Glasgow regenerative car covered 17,600 miles and showed an energy saving of 32%

While not an outstanding success, 305 was far from being a failure

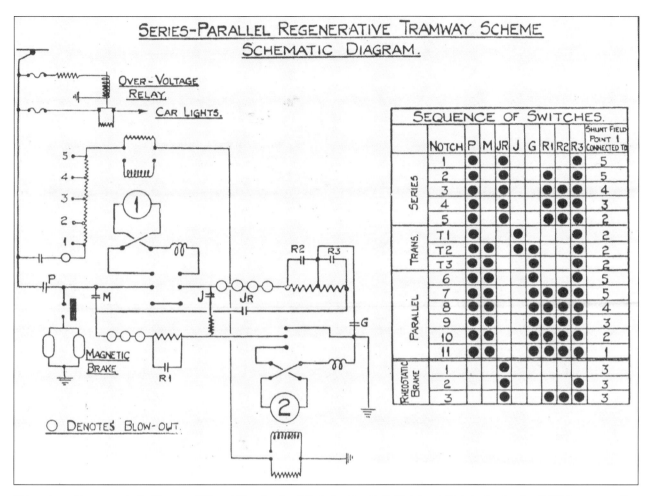

SERIES-PARALLEL REGENERATIVE TRAMWAY SCHEME
SCHEMATIC DIAGRAM.

OVER-VOLTAGE RELAY

CAR LIGHTS.

MAGNETIC BRAKE

○ DENOTES BLOW-OUT.

	NOTCH	P	M	JR	J	G	R1	R2	R3	SHUNT FIELD POINT 1 CONNECTED TO
SERIES	1	●		●					●	5
	2	●		●				●	●	5
	3	●		●			●	●		4
	4	●		●			●	●		3
	5	●		●			●	●	●	2
TRANS.	T1	●			●				●	2
	T2	●	●		●	●			●	2
	T3	●	●						●	2
PARALLEL	6	●	●				●	●		5
	7	●	●				●	●	●	5
	8	●	●				●	●	●	4
	9	●	●				●	●	●	3
	10	●	●				●	●	●	2
	11	●	●				●	●	●	1
RHEOSTATIC BRAKE	1		●							3
	2		●						●	3
	3		●				●	●	●	3

This schematic covers both Glasgow 305 and Manchester 420, being the G. H. Fletcher initial design for the M.T.T.A. regenerative equipment. Experimental in both fields of application, sundry modifications were inevitable and traceable in variations of this theme from time to time, and from site to site, of application. *(The Electric Railway, Bus & Tram Journal: 1931-September 18th: pp 32-145).*

Number 305 was Glasgow's first experimental (second phase) regenerative-braking tramcar, although it will be recalled that Glasgow did have two (first phase) JS Raworth equipments. 305 was also the first electro-pneumatic, remote controlled tramcar. Originally completed on 26th March 1909 at the Car Works at Coplawhill, it was fitted with its experimental equipment over the December 1932 – January 1933 period. Some ten months later electro-pneumatic remote control was installed. This was not an outstanding success as it was removed after another ten months, reverting to manual regenerative control. The modified electro-pneumatic control system was later most

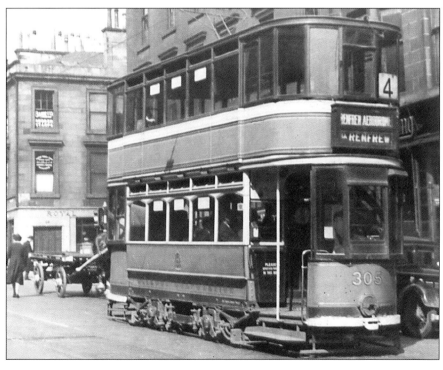

successfully incorporated into the subsequent design of the "Coronation" cars with only minor adjustments. Coplawhill Car Works records twice give the entry "OK40B Controllers" without dates of fitting and it has been taken, locally, to have been following the withdrawal of the remote control equipment on 23rd August, 1934. However, it is difficult to reconcile the addition of such purpose-built regenerative manual controllers with John Markham's suggestion that this particular model was not built until the following year, of 1935! A recorded change of motors at the end of May 1936 could indicate the removal of the regenerative control equipment and Car 305 returned to standard with series-wound propulsion. The other theory is that they had been withdrawn along with the electro-pneumatic equipment in 1934. There is no specific proof available either way. The car was obviously sufficiently successful to determine the Transport Committee to select regenerative equipment for the subsequent fleet of 40 manually-controlled regenerative tramcars. *(Photo: G Hunter)*

directions such as current consumption, extra speed of acceleration and deceleration in crowded streets and the extra safety in the briskly responsive equipment, that they went ahead with the ordering of 40 custom-built sets of regenerative equipments for such a fleet of these cars that had only been equalled in the Birmingham area during the first phase of regenerative braking in Britain.

It had been recorded on car 305 that a steady saving of 23% in current consumption, compared with other standard cars, regularly occurred with careful driving. As a single car, it would have been assured of a receptive overhead. This remarkable advantage was not reckoned as of so very much value on account of Glasgow's generation of her own traction current at Pinkston Power Station at such economical rates. However, very great stress was laid upon the further economy on brake-shoe and rail wear and tear.

However, very great stress was laid upon the further economy on brake-shoe and rail wear and tear

There was expense-less braking in normal service, amongst other attributes. This was despite the fact that 305 was a lone conversion and had to mingle constantly with non-regenerative cars all over the system which distracted from fully demonstrating its virtues.

Car 305's remote control equipment, which paved the way for the 'Coronation' cars of 1937, is recorded as being removed on 23rd August 1934. It may have been a demonstration set that Glasgow had operated on loan for trial purposes. At this time, it is almost certain that the regenerative equipment was likely to have been removed also, with a reversion to standard. The car worked on for a couple of decades with simple series equipment, before being scrapped at Elderslie Depot on 23rd/24th September 1954. The car had been observed working Service 32 (Provanmill – Elderslie) over August and September 1944 and on the Springburn – Charing Cross circular service 33 in September 1945 and February 1949, out of Possilpark Depot.

An interesting report dated 3rd May 1933 covers a month's field testing and investigation into the function of Car 305 as set against two, non-regenerative Standard cars, 94 and 108, on the Rouken Glen - Millerston – Bishopbriggs route. This covered the period 1st to 30th April, 1933. The author is indebted to Messrs BM Longworth and G Price for unearthing this material. It was out of service for part of Friday 7th April, for attention, and ran only 35 miles on that day. The average daily mileage otherwise had been 146.6 miles per day. The car clocked 4,397 miles during the month's trial period, at an average speed of 10.14 mph using 1.42 units of current per car-mile, as opposed to 2.14 units consumed by the investigation control cars thus showing an average percentage saving in current of 34% over cars 94 and 108 that were series-field control cars.

It was not to be wondered at that the Accountancy Department viewed with avarice that 34% figure of power saving regardless of the economy of the Tramway system's own Pinkston Power Station, as possibly projected into the then 1,060-odd

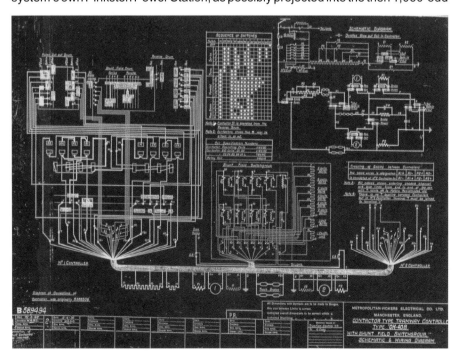

Contactor type Tramway Controller, Type MV-OK-40 B, with shunt field switchgroup – Schematic & Wiring Diagram for Glasgow Corporation Transport Experimental Regenerative Braking Tramcar No. 305. Note the date of origin 19th March 1933.

service passenger cars in the City's fleet. The Transport Committee fully agreed to a fleet of 40 cars to be fitted out with brand new 'state-of-the-art' regenerative equipment for extended in-service trial of this new type of control. Trial to what end? Certainly the ultimate concept could only have been the possible extension of the system to the 'Coronation' car fleet yet to materialise and still then under contemplation. Leeds' failure to project regenerative braking from their 1933 Middleton Bogie prototype (255) into the production fleet may have exerted reticence. Johannesburg had yet to pin its faith in regenerative control at this time. We shall never know now.

Apart from the admitted amazing saving in current, the increased service speed (inherent usually in regenerative braking) due to the rapid resumption of forward motion following recurrent braking in city centre hold-ups, may be observed from the following excerpted figures:-

	Car 305	**Car 94**	**Car 108**
Actual average speed	10.14mph	9.7mph	9.9mph
Over No. days tested	29½ days	28 days	29 days

Small as these figures seem, they were certainly not the full speeds available. They were averaged over the entire running of the cars on peripheral, as well as central lines. They were, however, very positive, and, had the motorman handling 305 been that bit more familiar with the car's characteristics over the very short period the regenerative braking (and electro-pneumatic control), were carried out (something time and experience would have provided), a different picture would have emerged.

The document from which the figures of the 305/94/108 regenerative braking trials were deduced has a history of its own having been salvaged from the mess of discarded 'rubbish' at the disposal of the former Tramways Head Office at 46 Bath Street in 1972. Although incomplete, it is of enormous transport archaeological significance and no other reports of these trials have been traced.

It was this short period of testing that prompted the Transport Committee to decide on the subsequent trial involving 40 cars thus vindicating the brief spell of assessment of 305. Hence this car was no failure despite the quick return to conventional control. Thus No. 305 remains a car outstanding in tramway history.

Car 305 was probably the prototype upon which Metro-Vick's order for Leeds 255, and 50 sets of equipment for Johannesburg was based.[17]

GLASGOW CORPORATION TRANSPORT

REGENERATIVE CAR TEST RESULTS

ROUKEN GLEN, MILLERSTON AND BISHOPBRIGGS ROUTE

CAR 305 REGENERATIVE

DATE		CAR MILES	TOTAL RUNNING TIME		S.C.H. SPEED M.P.H.	METER READINGS			UNITS USED MOTORING	UNITS GENERATED	UNITS USED PER CAR MILE MOTORING	UNITS GNR PER CAR MILE	UNITS USED FROM M & G	UNITS USED FROM TOTAL	UNITS USED PER CAR MILE FROM TOTAL
			Hrs	Mins		Motoring	Generating	Total							
Sat	April 1st	139	14	45	9.43	88995.7	33856.5	71095.0	287.6	70.9	2.06	.510	216.7	212.7	1.53
Sun	2nd	137	12	48	10.70	89244.0	33922.4	71276.3	248.3	65.9	1.81	.481	182.4	181.3	1.32
Mon	3rd	157	15	2	10.44	89512.1	33984.3	71480.7	268.1	61.9	1.70	.394	206.2	204.4	1.30
Tue	4th	150	14	35	10.30	89781.6	34051.7	71680.5	269.5	67.4	1.79	.449	202.1	199.8	1.33
Wed	5th	150	14	35	10.30	90105.2	34134.0	71919.2	323.6	82.3	2.15	.548	241.3	238.7	1.39
Thur	6th	150	14	35	10.30	90387.4	34207.3	72126.5	282.2	73.3	1.88	.488	208.9	207.3	1.38
Fri	7th	35	6	38	8.29	90491.4	34234.8	72203.5	104.0	27.5	1.89	.500	76.5	77.0	1.40
Sat	8th	160	15	43	10.18	90808.4	34316.1	72436.2	318.0	81.3	1.98	.508	235.7	232.7	1.45
Sun	9th	110	10	18	10.68	91012.9	34369.0	72586.3	204.5	52.9	1.85	.481	151.6	150.1	1.36
Mon	10th	130	14	35	10.30	91311.4	34446.2	72805.2	298.5	77.2	1.99	.514	221.3	218.9	1.46
Tue	11th	150	14	35	10.30	91607.2	34522.5	73022.2	295.8	76.3	1.97	.508	219.5	217.0	1.44
Wed	12th	150	14	35	10.30	91897.7	34597.5	73235.0	290.5	75.0	1.93	.500	215.5	212.8	1.41
Thur	13th	150	14	35	10.30	92180.7	34670.8	73442.1	283.0	73.3	1.88	.488	209.7	207.1	1.38
Fri	14th	150	14	35	10.30	92474.6	34747.6	73655.5	293.9	76.8	1.96	.512	217.1	213.4	1.42
Sat	15th	160	15	45	10.18	92783.2	34828.7	73882.1	306.6	81.1	1.93	.506	227.5	226.6	1.41
Sun	16th	153	13	37	11.23	93086.3	34913.1	74098.9	303.1	84.4	1.98	.551	218.7	216.8	1.41
Mon	17th	150	14	35	10.30	93353.7	34983.2	74291.8	267.4	70.1	1.78	.467	197.3	192.9	1.28
Tue	18th	150	14	35	10.30	93626.3	35052.4	74492.6	272.6	69.2	1.81	.461	203.4	200.8	1.33
Wed	19th	150	14	35	10.30	93923.3	35127.2	74712.2	297.0	74.8	1.98	.500	222.2	219.6	1.46
Thur	20th	150	14	35	10.30	94212.8	35202.4	74923.1	289.5	75.2	1.93	.501	214.3	210.9	1.40
Fri	21st	150	14	35	10.30	94513.1	35281.2	75142.7	300.3	78.8	2.00	.525	221.5	219.6	1.46
Sat	22nd	150	15	43	10.18	94837.3	35362.6	75383.2	324.2	81.4	2.02	.508	242.3	240.5	1.50
Sun	23rd	153	13	37	11.23	95134.3	35443.3	75596.6	297.0	80.7	1.94	.527	216.3	213.4	1.39
Mon	24th	150	14	35	10.30	95438.4	35525.6	75816.2	304.1	82.3	2.02	.548	221.8	219.6	1.46
Tue	25th	150	14	35	10.30	95734.6	35608.5	76025.7	296.2	82.9	1.97	.552	213.3	208.5	1.40
Wed	26th	150	14	35	10.30	96049.1	35690.9	76255.1	314.5	82.4	2.09	.550	232.1	229.4	1.53
Thur	27th	150	14	35	10.30	96345.9	35771.0	76470.8	296.8	80.1	1.98	.534	216.7	215.7	1.43
Fri	28th	150	14	35	10.30	96655.9	35858.1	76692.3	310.0	85.1	2.06	.567	224.9	221.5	1.47
Sat	29th	160	15	43	10.18	96963.4	35934.9	76918.0	307.5	78.8	1.92	.492	228.7	225.7	1.41
Sun	30th	153	13	37	11.23	97273.0	36021.7	77140.1	309.6	86.8	2.02	.567	222.8	222.1	1.45
	Total	4,397							8,564.9	2236.1			6,328.8	6,257.8	
	Average Values										1.94	.508			1.42

3rd May, 1933

GLASGOW CORPORATION TRANSPORT
REGENERATIVE CAR TEST RESULTS
ROUKEN GLEN, MILLERSTON, BISHOPBRIGGS ROUTE
CAR 94 (NON-REGENERATIVE)

DATE		CAR MILES	TOTAL RUNNING TIME		SCHEDULE SPEED M.P.H.	METER READING	UNITS USED	UNITS USED PER CAR MILE	% SAVED IN CURRENT BY CAR 305
	April		Hrs	Mins					
Sat	1st	161	16	3	10.03	31112.7	332.6	2.06	25.8%
Sun	2nd	152	13	25	11.20	31405.5	292.8	1.92	31.3%
Mon	3rd	175	17	25	10.05	31702.5	297.0	1.70	23.5%
Tue	4th	150	14	32	10.32	32027.5	325.0	2.16	38.5%
Wed	5th	150	14	32	10.32	32366.9	339.4	2.26	29.7%
Thur	6th	150	14	32	10.32	32685.2	318.3	2.12	34.9%
Fri	7th	150	14	32	10.32	33030.8	345.6	2.30	39.2%
Sat	8th	161	16	03	10.03	33369.1	338.3	2.10	31.0%
Sun	9th	152	13	25	11.43	33669.9	300.8	1.98	31.4%
Mon	10th	150	14	32	10.32	33998.1	328.2	2.19	33.4%
Tue	11th	150	14	32	10.32	34310.3	312.2	2.08	30.8%
Wed	12th	150	14	32	10.32	34630.1	319.8	2.13	33.8%
Thur	13th	150	14	32	10.32	34948.9	318.8	2.12	34.9%
Fri	14th	150	14	32	10.32	35266.6	317.7	2.11	32.7%
Sat	15th	161	16	03	10.03	35615.7	349.1	2.16	34.8%
Sun	16th	-	-	-	-	-		Not In Service	
Mon	17th	150	14	32	10.32	35934.7	319.0	2.12	39.7%
Tue	18th	150	14	32	10.32	36240.9	306.2	2.04	34.8%
Wed	19th	150	14	32	10.32	36590.9	350.0	2.33	37.4%
Thur	20th	150	14	32	10.32	36913.5	322.6	2.15	34.9%
Fri	21st	150	14	32	10.32	37241.1	327.6	2.18	33.0%
Sat	22nd	161	16	03	10.03	37572.2	331.1	2.05	26.9%
Sun	23rd	-	-	-	-	-		Not In Service	
Mon	24th	150	14	32	10.32	37930.8	358.6	2.39	39.0%
Tue	25th	150	14	32	10.32	38254.8	324.0	2.16	35.2%
Wed	26th	150	14	32	10.32	38595.0	340.2	2.26	32.3%
Thur	27th	150	14	32	10.32	38935.4	340.4	2.27	37.0%
Fri	28th	150	14	32	10.32	39286.4	351.0	2.34	37.3%
Sat	29th	161	16	03	10.03	39695.5	409.1	2.54	44.5%
Sun	30th	152	13	25	11.20	40017.9	322.4	2.12	31.6%
	Total	4,286	-	-		-	9,237.8	-	-
	Average Values		-		-			2.15	34.0%

46 Bath Street, Glasgow, C2
3rd May, 1933

GLASGOW CORPORATION TRANSPORT
REGENERATIVE CAR TEST RESULTS
ROUKEN GLEN, MILLERSTON, BISHOPBRIGGS ROUTE
CAR 108 (NON-REGENERATIVE)

DATE		CAR MILES	TOTAL RUNNING TIME		SCHEDULE SPEED M.P.H.	METER READING	UNITS USED	UNITS USED PER CAR MILE	% SAVED IN CURRENT BY CAR 305
	April		Hrs	Mins					
Sat	1st	159	16	10	9.83	26149.8	335.0	2.10	27.2%
Sun	2nd	-	-	-	-	-		Car Not In Service	
Mon	3rd	180	17	11	10.48	26492.0	342.2	1.90	31.6%
Tue	4th	160	15	52	10.08	26821.6	329.6	2.06	35.5%
Wed	5th	160	15	52	10.08	27156.4	334.8	2.09	24.0%
Thur	6th	160	15	52	10.08	27497.8	341.4	2.13	35.3%
Fri	7th	160	15	52	10.08	27836.8	339.0	2.12	34.0%
Sat	8th	159	16	10	9.83	28157.9	321.1	2.02	28.3%
Sun	9th	153	13	31	11.33	28500.8	342.9	2.24	39.3%
Mon	10th	160	15	52	10.08	28864.9	364.1	2.27	35.7%
Tue	11th	160	15	52	10.08	29212.7	347.8	2.17	33.7%
Wed	12th	160	15	52	10.08	29580.3	367.6	2.30	38.7%
Thur	13th	160	15	52	10.08	29937.6	357.3	2.23	38.2%
Fri	14th	160	15	52	10.08	30293.5	355.9	2.22	36.0%
Sat	15th	159	16	10	9.83	30617.2	323.7	2.03	30.6%
Sun	16th	153	13	31	11.33	30943.5	326.3	2.13	33.8%
Mon	17th	160	15	52	10.08	31289.7	346.2	2.16	40.8%
Tue	18th	160	15	52	10.08	31609.2	319.5	2.00	33.5%
Wed	19th	126	13	22	9.43	31887.3	278.1	2.20	33.7%
Thur	20th	160	15	52	10.08	32232.0	344.7	2.15	34.9%
Fri	21st	160	15	52	10.08	32574.9	342.9	2.14	31.8%
Sat	22nd	159	16	10	9.83	32923.1	348.2	2.19	31.5%
Sun	23rd	152	13	25	11.20	33256.0	332.9	2.19	36.6%
Mon	24th	160	15	52	10.08	33623.2	367.2	2.30	36.6%
Tue	25th	160	15	52	10.08	33985.9	362.7	2.26	36.1%
Wed	26th	160	15	52	10.08	34343.9	358.0	2.23	31.4%
Thur	27th	160	15	52	10.08	34696.8	352.9	2.20	35.0%
Fri	28th	160	15	52	10.08	35065.5	368.7	2.30	36.1%
Sat	29th	159	16	10	10.18	35408.5	343.0	2.15	34.5%
Sun	30th	153	13	31	11.33	35756.0	347.5	2.27	36.2%
	Total	4,592	-	-		-	9,941.2	-	-
	Average Values		-		-			2.16	34.3%

3rd May, 1933

THE EDINBURGH REGENERATIVE BRAKING EXPERIMENTAL CAR – 203 OF 1933

Edinburgh, like Glasgow, first experimented with a single car – 203. This car was new, turned out of the Shrubhill Works in February 1933. It was the second car to carry that number. It was equipped with the Metropolitan-Vickers system of regenerative braking control involving MV101BC compound-wound 50 hp motors and, possibly, British Thomson-Houston B 510 controllers appropriately modified by Metro-Vick. There is no formal evidence as to which type of controllers Edinburgh used for this experiment.

It is unlikely that the M-V Type OK-40B regenerative controllers had been completed by this time and M-V would have been required to modify a pair of Edinburgh's standard controllers for this experiment.[18] The tests were promulgated under the aegis of FA Fitzpayne who had come to the Edinburgh system from the General Manager's position with the Leith Corporation Tramways, becoming the Edinburgh General Manager from January 1929. GH Fletcher, Chief Electrical Engineer and Manager of M-V's Sheffield Traction Motor Department, was prominent in overseeing the practical performance of the experiment.

This was equal to a saving of £65 per car per year

Edinburgh's Transport Sub-Committee had authorised the purchase of one set of regenerative control equipment on 15th July 1932. This was fitted to Car 203 and put on trial on the Portobello route for several months. There the satisfactory performance showed an average saving in current consumption of approximately 30% comprising some 0.36 units of consumption per car-mile. This was equal to a saving of £65 per car per year and there was also the added benefit of enhanced acceleration and speed. This represented an overall increase in performance at reduced cost. It led the Public Utilities Committee to authorise the purchase of eleven more sets of the equipment on 22nd December 1933. The Portobello – Levenhall route was the only one supplied through rotary converters and not mercury-arc rectifiers. This determined the transfer of the regenerative fleet to Portobello Depot.

Speed was the Magazine of the Edinburgh Transport Department. The spring 1933 issue, although unfortunately short in detail of technical information, does give that 'The controller has only three resistance notches, the remaining eight all being running notches, so that in addition to the economy from regeneration, a further economy is accomplished by the elimination of resistance notches'.

Passengers very quickly complained of the jerky characteristics of the equipment

Despite all these remarkable attributes there were drawbacks to the regenerative system. Passengers very quickly complained of the jerky characteristics of the equipment. This, together with the constant 'lamp popping' phenomenon, gave

Edinburgh's prototype regenerative braking tramcar – 203, of 1933 at Joppa on 19th June of that year. *(Photo: D.L.G.Hunter, courtesy AW Brotchie).*

them grist to the mill of complaint. Metro-Vick's GH Fletcher investigated this over-voltage phenomenon and assessed variation in voltage due to regeneration had never been more than 2% above the normal motor voltages. This was an infinitesimal figure, dependent upon the amperage carried. It could almost have been measured in the number of light bulbs fused! Fletcher's tests must have been done when the line was receptive!

In the April 1934 number of the Metropolitan-Vickers Gazette – reprinted from 'The Electric Railway, Bus and Tram Journal' (pages 378-381), Eric R. L. Fitzpayne, while working in Edinburgh, wrote up the 'Edinburgh Speed Trials' on the practical assessment of value to be obtained by increasing the speed of tramcars.[19] Three Edinburgh tramcars, of the same batch and with standard MV101 motors, were used in the trial, their motors being altered in the following fashion:-

Car 162   High-speed series-wound MV101F: speed increased by reducing both field and armature turns.
Car 203   Regenerative braking: MV101BC compound windings instead of field-tapping otherwise the same as Car 206.
Car 206   Tapped fields: MV101G (local designation): Tapped fields giving two steps of field weakening, no change in armature turns.

All cars weighed 12 tons 7 cwt empty, and 12 tons 10 cwt with motorman, conductor and observers. The gear ratio was standard at 14/61 and the motors were wound with Class 'B' insulation. Not only were the standard MV101 motors modified as above but also, in order to gain a true comparison of the three systems, the motors were modified to give a similar performance on the maximum speed position on the controllers. This brought the top speed curves for the three motors sufficiently close to permit a fair comparison when working to the same schedule. The one-hour rating of each of the motors was:-

	HP	RPM	AMPS
MV101 (standard)	47	800	72
MV101F (high speed, series wound)	69	1220	105
MV101BC (regenerative control)	47	820	74
MV101G (tapped fields)	47	765	74

The line voltage throughout was 550 Volts.

The main consideration in speeding up was naturally the cost of increased energy consumption. This was assessed in the trial as was ascertaining which of the three methods of speeding up was the most economical. In the case of the Regenerative car, this involved determining how much energy was used up in the resistors in starting and in the wheels and track brakes in stopping that could be avoided in practice by the use of field diverting or by regenerative braking. To increase comparability the tests were made over fairly level routing so that the special advantages of the regenerating equipment on hilly routes were eliminated.

The tests involved scheduled running on both a single round trip and of two round trips. With varying traffic conditions this was made difficult as estimation of necessary adjustments to theoretical times given on the record sheets raised observer failures. Power consumption in units was shown in terms of the double round-trip and the readings were corrected for voltage and adjusted to allow for meter errors and other discrepancies that inevitably arose.

Car 206 saved 9.05% power consumption over Car 162 with high-speed straight series motors. Car 203 (the Regen car) saved 32.8% in power consumption over Car 162 and 26.1% over Car 206 in terms of raw, non-adjusted figures, showing that the Regenerative car was outstandingly the most economical as far as practical running was concerned.

the Regenerative car was outstandingly the most economical as far as practical running was concerned.

The average acceleration from rest, up to 20 mph was 2.0 mph per second, while the average braking retardation ran at 1.5 mph per second.

Through the courtesy of FA Fitzpayne, particulars had been furnished of the performance of the regenerative control equipment purchased for experimental purposes by the Edinburgh Corporation. This set had covered over 48,500 miles and had regenerated more than 27% of the energy taken for motoring. It had been run in comparison with a number of standard cars in regular service. Comparison of the records for these days when the experimental car was in service with one particular standard car shows that for a mileage of over 32,000 miles the saving in energy was over 18% of the standard car's consumption. On account of this favourable performance the Edinburgh Corporation ordered the eleven further sets of regenerative equipment at a cost of £2,644 for the conversion of cars for a route where the cost of electrical energy was unduly high.[20]

Speed, of 1933, shall have the last word on 203:-

'When the controller is moved back from the full parallel position, this field (the shunt field) is strengthened, the voltage across the field* rises and current is returned to the line. The return of current to the line checks the speed of the car and thus, whenever the controller is moved back into the off position, the car automatically brakes, the controller handle now acting much as the accelerator of a motor car, each notch of the controller being equivalent to a certain predetermined speed. Thus, if a driver wishes to descend a hill at say 10 mph, he finds the notch for this speed, and the motors do the rest.

The electric car equipped with air brakes, rheostatic brake and mechanical brake is recognised as the safest vehicle on the road, but this car, with its additional regenerative brake is even safer. The controller has only three resistance notches, the remaining eight notches all being running notches, so that in addition to the economy of regeneration, a further economy is accomplished by this elimination of the resistance notches.'

'Blowing your own trumpet isn't the best way of facing the music.'

* It is, in fact, the rise in armature voltage that drives power back into the line.

THE LEEDS EXPERIMENTAL REGENERATIVE BRAKING CAR — 255 OF 1933.

The only British 4-motor equal-wheel bogie regenerative car to enter service . . .

The edition of *Modern Transport* for 3rd June 1933 described the Leeds City Tramways and Transport Department's experimental regenerative braking tramcar, 255. This was an elegant, well-balanced design of a totally enclosed, double-bogie vehicle. It was equipped with Metropolitan-Vickers 4-motor, electro-pneumatic remote controlled regenerative braking equipment. The builders were The Brush Electrical Engineering Company of Loughborough. Number 255 was designed for the hilly Middleton Light Railway branch in the southern suburbs of the city and entered service in July of that year: the first, and as it transpired, the only British 4-motor equal-wheel bogie regenerative car to enter service.

The four compound-wound Metropolitan-Vickers Type 109CZ regenerative motors developed 30 hp each, at 275 Volts on a one hour rating, being connected two in permanent series. They were mounted on Maley & Taunton (Wednesbury) swing-link trucks – a new design.

The car's main dimensions were 36ft (10,973mm) in overall length, over the fenders and 6ft-11in (2,108mm) in width. There was a low overbridge on the line and the height was restricted to 14ft-9½in (4,509mm). The body length over the corner pillars was 23ft-6in (7,163mm). The bogies incorporated 27in (685mm) diameter wheels set at a wheelbase of 4ft-9in (1,403.5mm) with pivotal centres of 14ft-0in (4,267mm). The weight of the trucks, motors, air and magnetic brakes came to 6 tons 10 cwt (6,604 kg).

Seating accommodation was provided for 70, 30 in the lower saloon and 40 above, all on transverse seats with the exception of a longitudinal seat at each end of the lower saloon opposite the staircases.

Number 255 rejoiced in the modernity of enclosed motormen's cabins which included a seat for the motorman with all controls. Also later included was electro-pneumatic contactor equipment at one end, while the other cabin housed a compressor. Pity the poor motorman in summer weather! The original controllers were MV Type OK42B.

Leading the way, indeed in world tramway equipment and passenger comfort which Johannesburg, Liverpool and Glasgow would follow, 255 was unfortunately

Top: Leeds 255 in very early days, still with twin trolleys. A detail of the end view.
(Photo: John D Markham collection)

Lower left: General view of the motorman's cabin on car 255. It is the then-current blue livery.
(Photo: John D Markham collection)

A close-up view of 255's cabin showing the electro-pneumatic contactor cabinet on the right, together with the smaller remote controller that this allowed, centrally installed. The eight small contactors at the top were for the shunt field control while the electro-pneumatic contactors at lower right served the resistor notches (R1-R4) and transition (G, TR, JR and M).
(Photo: John D Markham collection)

not a success – again possibly from an excess of untried experimental equipment. It was in January 1934 when the electro-pneumatic regenerative equipment was commissioned.[21] Another drawback in what was a pleasing design was the very narrowness of the access steps and platform entrances which were the cause of a degree of congestion. This was later modified in the production batch of virtually identical vehicles that followed.

In November 1933, 255 was placed in regular service on the Middleton Light Railway. The Leeds General Manager, W Vane Morland, had kindly furnished details of the car and of the results obtained from it. 255 was unique in that it was driven by four regenerative motors in two equal-wheeled bogies. The total horsepower on a one-hour rating was 120. The original control gear was cam-operated OK42-B with electro-magnetic contactors for shunt field control. This gave frequent trouble hence replacement with electro-pneumatic remote control the following January. From that time the car operated without trouble for 200 miles per day, seven days per week.[22]

Over approximately 20,000 miles the maintenance costs of the regenerative equipment had been negligible

The Middleton Light Railway was some 4½ miles long. Around one third was through public thoroughfares and the remainder was on sleeper track with a long steady gradient. An overall schedule speed of 12 mph was maintained but the average speed over the reserved track was over 15 mph. Over approximately 20,000 miles the maintenance costs of the regenerative equipment had been negligible. At the same time wear and tear on brake blocks had been considerably reduced. On account of 255 being the only one of its type it had not been possible to obtain comparative figures of energy consumption from a non-regenerative car of similar design. However, a test run under normal conditions had shown that the energy returned to the line had been 31.5% of the energy taken while motoring.[23]

Number 255's MV109CZ motors were compound having four main and four interpoles. Many previous tramway motors used the axle housing, recessed into the shell of the motor, as the fourth (unwound) interpole. The armatures, 10in diameter and 6.7in core length, had 29 slots with two turns per slot. Skefco roller bearings were provided at each end and the total weight was 250 lbs.

The car was demonstrated to the press and entered service with two OK42B camshaft type, direct controllers and the four MV109CZ motors mentioned. The MV109CZ motor was an MV109Z rewound with compound fields for regeneration. Whilst mechanically similar, armature and field windings were different. MV109Z motors were used on the second batch of London HR/2 cars and were an axle-hung version of the MV109 motors specified for the earlier batch.

In November 1933, Car 255 was converted, at Metro-Vick's expense, to electro-pneumatic control.

Broadside maker's view of car 255.
(Photo: Courtesy, National Tramway Museum archive)

According to Jim Soper, Car 255 was converted, in 1952, at the behest of the Leeds Engineer, VJ Matterface, from regen. to non-regen. by the substitution of four type MV109GZ motors Leeds had ordered in 1951. The MV data sheet for these motors states that they were for a prototype single-deck car, and these were the motors Installed into new tramcar 601 and indeed motors of this type were fitted to that car. As two of its motors are stored at Clay Cross, the storehouse of the National Tramway Museum, it should be easy to determine from their serial numbers that they formed part of the 1951 order.

John Markham is convinced of an alternative possibility. Matterface was ex-London Transport, and would have been aware that the early HR/2 cars to be withdrawn were those from the second batch that had no trolleys. They had MV109Z series motors, with axle suspension. It would have been perfectly feasible that the four MV109CZ regenerative motors from 255 were removed, despatched to London and surreptitiously exchanged for four good series-wound MV109Zs taken from an HR/2 car destined for scrap. The copper value would be 34 lbs per motor in London's favour. It would explain Matterface's reluctance to say just what type of motors were fitted to 255 in its latter days. Such an exchange could have been through an 'Old Boy' network, and not documented. The return transport to and from London would be at Leeds' expense but would probably cost less than reworking the regenerative motors to become series machines.

After conversion, the resistance notching for car 255 was different from the other Middleton Bogie cars. (A handwritten note on a print of the schematic diagram held by this author mentions this suggesting a motor characteristic different from that of the GEC machines of the main batch of Middleton Bogie cars.)[24]

The compound-wound regenerative motors complete except for gears and gearcases, weighed 1003 lbs. They were rated, on a one hour rating at 120°C, with a current value of 275 Volts, at 95 Amps in the main circuit and of 4 Amps in the shunt circuit producing 30 shaft horsepower and 1,040 rpm per motor. Thus the total one-hour rating of the car came to 120 h.p. On the continuous rating at 105°C, the figures were, per motor, 24 shaft hp, using 75 Amps, at the same voltage, and at a speed of 1,120 rpm. The shaft efficiency in both cases was 84%. The shunt Amperage was 4 Amps at the one-hour rating and 3.5 amps under continuous rating. It is probably that the regenerative control was replaced by standard equipment in 1951 using MV109Z type motors ex-London HR/2 (series II) cars.[25]

The production batch of 16 Leeds Middleton Bogies (256-271) had G.E.C. W/T 181 A motors which had a one-hour rating of 35 hp, at 275 Volts, being four-motor equipments, series wound.[26] Further information on the Leeds cars appears in Appendix 4.

THE EXTENSION OF LONDON'S REGENERATIVE BRAKING PROGRAMME, 1933

The previous London experiment with the MTTA equipment from Manchester, on LCC Class E/1 car 779, had been inadequate due to the inability of the Manchester motors to cope with a heavier car. At 37.5 hp on a one-hour rating, a couple of these under-powered prime movers still showed such a significant saving in current consumption, as well as in brake shoe and rail wear, that the LCC authorities were deeply impressed. Out of this came the decision to extend the research further into a range of some four sets of modern, bespoke regenerative equipment by different makers with the obvious anticipation of widespread application, should these results hold good.

The LCCT therefore ordered the following equipment and had them fitted to the following Class E/1 cars. Meanwhile the 1933 amalgamation of all of London's tramways, buses, trolleybuses and underground railways into the massive LPTB, or London Passenger Transport Board, matured.

781 G.E.C. W/T 293 A motors at 55 h.p. rating	}	
782 MV 114 CR motors at approximately 55 h.p. rating	}	Regenerative
783 E.E.304 motors at approximately 56 h.p. rating	}	Braking
784 Crompton-Parkinson C150B motors (around 55 h.p)	}	machines
785 E.E.301A Regulated Field (non-Regen) at 56 h.p.	}	Regulated
786 Crompton-Parkinson C150B motors 55-60 h.p.	}	Field Control
787 G.E.C. W/T 291 A motors at 65 h.p. rating	}	(weak field)
788 B.T-H 509 S1 motors (probably 50-55 h.p.)	}	machines [28]

Well planned though it may have been for an assessment of the best available state-of-the-art regenerative braking motor on the market, the interest had petered out virtually before delivery as the LCC Tramways had been absorbed into the LPTB on 1st July 1933. This was despite two of the L.C.C. senior men Messrs TE Thomas (later Sir Theodore) being appointed as General Manager Road Transport, and GF Sinclair as Rolling Stock Engineer to Deputy General Manager (Road Services). Effectively they did not lose rank but the new combine's avowed intent had been the entire elimination of their inherited tramways and the quicker the better. The new motors were delivered but the interest flagged due to the intensive reorganisation within the enormous new transport machine and nothing more was heard of it. It simply faded out of existence, unremarked.[29]

The LCC clearly wanted to make comparisons with the cheaper and less technically risky alternatives to regen . . .

Of these new motors, four (Cars 781-784) were regenerative braking machines while the other four (Cars 785-788) had regulated field control.[30]

The LCC clearly wanted to make comparisons with the cheaper and less technically risky alternatives to regen. So they used machines designed for wide-range field control by which the number of economical running notches could be increased and less of the accelerating cycle be dependent on power dissipation in the starting resistances, without introducing the losses associated with shunt field excitation that regen required.

MANCHESTER'S SECOND PHASE OF REGENERATIVE BRAKING EQUIPMENT – CAR 420 OF MAY 1933.

London returned the MTTA equipment to Manchester in May 1933 when it was again fitted to Car 420. A further 20,000 miles were run in test against several of the same class of series-wound vehicles. These showed savings in power consumption of 23.5%, over all. The M-V controllers had been modified in London and required more attention back in Manchester to equate with the experimental equipment being fitted to Glasgow's 305, by Mr Fletcher.

A series of correspondence from Metro-Vic at that time has come down through the hands of F Philip Groves to illustrate the type of recurrent problems inherent in the regenerative equipments of this Second Phase of their application to tramcars.[31] The much advanced technique of the 1930s was, even then, short of adequacy in extracting the full benefit of the regenerative process. This had to wait until the advent of solid-state electronic control for its maturity, as indeed the letters do show!

The third letter, dated July 1933 and reproduced here, was when the MTTA equipment had been returned to Manchester from London and established in Pilcher Car 420 once again. It mentions the need to bring it up-to-date with Mr Fletcher's continuing experimentation and improvement all over the country. The senior engineer signing the letter was HK Ramsden, first assistant to Mr Brooks at the Trafford Park Works of Metropolitan-Vickers.

The MTTA Annual Report of 1934-35 highlights the results with the Association's regenerative control equipment, both in Manchester and in London, and mentions also, similar regenerative experience in Edinburgh, Glasgow and Leeds. It was of the opinion that this equipment, purchased in 1931, had served its purpose well and had provided all Members with several reports on its performance on two large tramway systems. These had been confirmed independently by the experience of three other Members who had benefited by the advantages of improvements resulting from the Association's set. A number of such tramcars were then in regular service in Edinburgh, Glasgow, Halifax and Leeds and the system had been increasingly incorporated in trolleybus developments. It was agreed that there appeared to be no further need to pursue the experiment with the original apparatus and the Association sold the equipment to the Manchester Corporation Transport Department, outright.

The last of the letters, previously mentioned, that dated 5th November 1934, reveals a history of failing equipment. This included a failed armature, shunt field fusing, earthing of leads, controller faults, further armature faults and finally a badly damaged motor requiring extensive (and expensive) repairs. As there is no mention of further trouble or adjustment with the ageing MTTA equipment it would appear not unreasonable to conclude that Car 420 had, about this time, been stripped of its regenerative equipment and returned to series-wound machinery, the MTTA equipment being no longer required. Disposal of the equipment would then have

It was noticed that the driver made a good deal of use of the second series notch for regenerative braking. This, of course, is a resistance notch and some of the energy of regeneration is thus being lost to the resistance.

The reason for the driver's liking for this notch is that the shunt current on the resistance notches (1 and 2) is greater than on the first economical notch (3), and the minimum regenerative speed is thus somewhat lower on notch 2 than on notch 3.

The M.T.A. equipment is the only one where this condition obtains, and it is probably that the economy is slightly affected thereby.

It is intended to put the car into service on the Manchester – Oldham route.

(This letter is reproduced courtesy F.P.Groves)

File 929250

July, 1933.

M.T.A. REGENERATIVE TRAMCAR EQUIPMENT.

This equipment having been returned from L.C.C. Tramways to Manchester Corporation Tramways, it was re-erected on Car No. 420 for service use in Manchester.

Since it was last running in Manchester the control scheme has been modified to line up with the Glasgow equipment, i.e., the new transition and rheostatic brake scheme, with contactors for the shunt field control.

At the request of the Manchester Tramways the writer, in company with Mr. Collis, visited the Hyde Road depot to check the operation of the equipment before it was passed into service.

A preliminary run was made, and following certain minor adjustments a second run, during which the performance was found to be quite satisfactory.

It was noticed that shunt contactor S.6 did not clear its arc satisfactorily when switching off from last rheostatic brake notch where it is closed to stiffen up the shunt field at the end of braking. The blow-out coil had been repaired, presumably in London, so it has evidently given some trouble previously.

As the strong shunt field on the last notches seems to be quite unnecessary, the controller contacts which operate S6 in this position were removed and S.6 contactor will thus be relieved.

The overvoltage relay had to be loaded to suit the L.C.C. voltages and so the weights were removed. The average voltage on the Manchester system appears to be 520 to 540, and in spite of severe regenerative braking on the trial runs the voltage did not exceed 580, which was insufficient to lift the overvoltage relay even without the weights.

The lighting circuits consist of 5 lamps of 105 volts each, i.e., 525 volts total. The protection given by the relay is, therefore, not good, as when hot the pick up voltage will exceed 600 volts.

MEMORANDUM OF VISIT TO MANCHESTER CORPORATION TRAMWAYS

5th November 1934.

In response to the request Mr. Blackburn, Rolling Stock Engineer of the Manchester Tramways made to Mr. Brooks, I visited the Hyde Road Depot with Mr. Collis on the 8th inst. to investigate the troubles which had been experienced on the regenerative equipment.

The history of this equipment since it was returned from the L.P.T.B. is, apparently, as follows :-

July 1933.   Put into service.

Sept.1933   Armature 14896 failed and was sent to Sheffield for rewinding.

Jan. 12,1934   Trial run after repairs.

Jan. 13   In service fuses blew. This fault was traced to intermittent open circuit on the shunt field connections in the control equipment.

Jan. 26.   TR lead earthed in the controller.

Feb. 1st   In service again.

Feb. 24.   Fuses blown. The equipment was examined by Mr. Collis and defect found in the Controller. Put back into service.

June 8th   2 Shunt resistances burnt out.

July 24th   Both armatures earthed. Repaired by Manchester Trams.

Oct. 24th   In service again.

Oct. 25th   Brought in burnt axle box.

Oct. 26th   No 1 Motor badly damaged due to band coming off.

that we visited Hyde Road. Both motors had been removed from the car and dismantled.

a. Damaged Motor.

The back end core band and the core band adjacent to it had come off this machine and had badly damaged the insulation both on the armature and on the field coils. Several coils of the armature had been short-circuited but there was only a slight amount of burning on the conductors. Several coils showed signs of having been overheated. In one slot the commutators had lifted and subsequently been torn out for about 3" by contact with the poles. There were two points on the commutator where heavy arcing had taken place and where the bars were shorted together.

The rest of the commutator was in good condition and had obviously been skimmed only a short time previously. The under cutting was deep but the micas had not been really well cleared out and there was a slight burring on certain of the bars and the mica had not been cut away from the edges of certain bars.

On the shunt field coils, the band had cut right through to the copper on three coils.

Mr. MacMahon, the Depot Superintendent said that he went out with the car on its trials on October 24th and everything appeared to be in order. There was no sparking at the commutators. Two days later the motor failed suddenly in service.

b. Undamaged Motor.

This armature had been sent to the Tramways Winding Shop to have a new band put on in view of the suspected trouble on the other machine.

The machine did not appear to have been overheated and the commutator surface was in very good condition having only recently been skimmed. As on the other machine the undercutting of the mica was deep but the finish of the micas left much to be desired.

I am inclined to agree with Mr. Blackburn that the cause of the trouble on No. 1 motor was the band (which had been put on by the Tramways) coming loose but the possibility that the trouble started as a short

between comm. bars cannot be entirely ruled out.

With regard to the repair of the machine Mr. Blackburn is considering whether he will order new field coils from us or attempt to wind new shunt coils in the Tramways Shops. It will almost certainly be necessary for him to order a number of new armature coils and in this connection he specially asked that we would do everything possible to expedite delivery.

(This letter is reproduced courtesy F.P.Groves)

been to scrap. During its second trial period in Manchester a few minor replacements had been necessary to the shunt field resistors and controller contactor springs.[32]

It has not been found possible to locate a reasonable photograph of the Manchester Car 420 *per se* – except as it its much later days as Aberdeen 41. Similarly no photograph of LCC 779 has come to light, although an attempt to reproduce the appearance of the car appears on the cover.

Subsequent to the return of the MTTA set to Manchester in May 1933, and its re-assembly in Manchester Car 420, it operated another 19,907 miles showing a saving in energy consumption of over 23.5% over the standard controls. During this period a few minor replacements had been necessary to the shunt field resistors and the contactor springs.[33]

EDINBURGH: THE 'PRODUCTION' BATCH OF REGENERATIVE CARS OF 1934–35

In the Annual Reports for the years ending 28th May 1935 and 1936 a maintained economy in current consumption of 0.36 units per car mile with the regenerative car 203 had been obtained. In addition further economy had been achieved through a marked reduction in the wear of rails, brake shoes and tyres. This was while the car had been working on the Musselburgh route. As has been seen, the Edinburgh

Edinburgh's production batch of eleven regenerative braking tramcars is epitomised by Car 119, taken at Joppa on 16th April 1938. Route 22 was a short working of the 21 to Musselburgh Town Hall only.
(Photo: The late D.L.G.Hunter, courtesy Alan W.Brotchie)

Edinburgh Regen car 149 alongside standard car 359 in1938.
(Photo: STTS collection)

Public Utilities Committee had been moved to authorise purchase of a further eleven sets of equipment from Metropolitan-Vickers on 22nd December, 1933.

A start was made in June 1934 to fit the eleven additional equipments and proceeded at a rate of approximately one car a month. The last one, 369, was completed in April 1935. All were existing cars, modified, and not built new with regenerative units. All were despatched to service as regenerative braking cars in the following order: 179; 65; 167; 161; 168; 166; 119; 149; 207; 368 and 369.

There is less detail as to why, in the face of all the reported attributes, that they were discontinued, but the jerky characteristics and the 'lamp popping' phenomenon also experienced in Glasgow, were not popular with the travelling public and the maintenance costs were not light. The economies presented a good case for retention but the equipment was gradually removed from the cars starting with 179 in February 1939. The last two, 161 and 168 retained their regenerative equipment until May 1940. All these cars had been taken from Portobello Depot and replaced by 'Pickerings' numbered 250-259, though some later did reappear at Portobello Depot.

a major flash-over occurred in the local rotary-converter substation paralysing the entire area for quite some time

Nevertheless, the tale is told of the first car out of Portobello Depot, early one morning – it would have been a regenerative car. It had to brake violently for some reason or another whereupon a major flash-over occurred in the local rotary-converter substation paralysing the entire area for quite some time while repairs were effected. It was well recognised that of the available converters of the period, rotaries were the least able to cope with regenerative voltage surges.

The problem probably occurred due to the decompounding of the fields by excessive reverse current injection on the dc side.

They had the standard regenerative braking equipment of Car 203 comprising MV101BC motors and, probably, BT-H. OK40B controllers, so were little different from the original experimental car.[34] The standard Edinburgh controller was the BT-H. B510 and the photograph that appeared in the MV records at Trafford Park of one such controller modified for regen use may point to its use in Scotland's capital.

THE HALIFAX REGENERATIVE BRAKING EXPERIMENTAL CAR 17 – 1934

Why Halifax should have joined, in 1934, the bandwagon of interest in the remarkable savings to be made from regenerative braking remains an enigma.

From 1933 onwards, the tram routes of this most hilly and windswept Yorkshire Pennine town were being systematically and fairly rapidly replaced by motor buses.

It is reasonably obvious that this had been due to the rising costs of electric power. Presumably even in the four years or so before the ultimate demise of the system, a small fleet of regenerative braking tramcars must have recouped the costs of replacement motor buses sufficiently to have made them an economic investment. After all, the outlay consisted only of regenerative equipments to be fitted to their existing motors within existing tramcars as against completely new motor buses. The total cost of the regen equipment was around £300 per set.

Yet in October 1934 Halifax's second electric car bearing the number 17, was fitted with a pair of British Thomson-Houston Type GE58 traction motors, rewound with compound fields, for regenerative braking purposes by Metropolitan Vickers (both by then owned by AEI or Associated Electrical Industries Limited of Trafford Park, Manchester). The experiment was so successful that right away five more equipments were ordered from M-V, in December of that year, immediately following Car 17's short period of running on the Causeway Foot route between October and 11th December 1934. During that time 17 lived up to expectations saving 1.05 units of current per car-mile. This was equivalent to £146 per annum in those days. The slipper brake shoes, also, lasted for 30,000 miles as against the nominal 4,000 miles in Halifax. The total outlay per car was £316-6-0 (£316-30p).

The five other new equipments were promptly apportioned to the existing double decker cars 116; 118; 119; 121 and 126 in December 1934. No time was lost in extracting the economy in current consumption inherent in the system.

Car 17(ii) was built new by Halifax Corporation in 1921 to replace the first bearer of that number. It was the first of ten new cars of the 'Hall' class named after Ben Hall who had become General Manager upon the retirement of JW Galloway in that year. Number 17 was originally a double-deck, open top, 4-window saloon car but fitted with a 3-window top cover with full length balcony canopies in October 1934

This view of Halifax tramcar 116 at Hebden Bridge in 1931 confirms the hilly nature of the terrain traversed by the trams. *(Photo: SL Smith, courtesy BJ Cross)*

Halifax 17, awaiting scrapping. The late Maurice O'Connor managed to record Halifax's experiment with the Second Phase regenerative braking shortly before 17's demise. The short top deck enclosure gave some top deck protection without too much side exposure to the buffeting side-winds on Halifax's exposed hilly routes. *(Courtesy: The National Tramway Museum Archive)*

A typical GE58 motor as fitted into Halifax 17 during its regenerative braking years. This one was installed in a Paisley District Tramways car inherited by Glasgow Corporation Tramways.
(Photo: Struan J.T. Robertson)

at the same time as the regenerative equipment was being installed. The car was 25ft 0in long over all with a body length of 16ft 0in over the corner pillars. The width was 6ft 6in and the height 14ft in to the trolley-pole head on the open top deck of its earlier days. Seating for 63 passengers was provided; 22 in the lower saloon and 41 above. The truck consisted of a modified Peckham Cantilever fabricated type with roller bearing axles, 30in wheels and 6ft 6in wheelbase, apparently carrying two Metropolitan-Vickers 40 hp traction motors. This would suggest that the GE58 motors converted to shunt fields for regenerative purposes must have come from stock, possibly from a previously scrapped car, for convenience. It is of interest that such an old type of motor should lend itself to the necessary modern rewinding, yet not become prone to overheating.[33] Later conversions involved DK30/1L 50 hp equipment. February 1939 saw the closing of the Halifax's system and the last service withdrawn was the route from Ward's End to Mason's Green and to Skircoat Depot.[35,36,37]

THE PRODUCTION BATCH OF FORTY GLASGOW REGENERATIVE BRAKING CARS OF 1934–35

Following experience with the experimental car 305 the Town Council of Glasgow agreed the motion of their Transport Committee to order no less than forty sets of, by then, custom-built regenerative braking equipments from the British Thomson-Houston firm of Rugby. These forty sets of regen equipment were earmarked for the last cars of the ultimate conversions from 'Semi-High Speed' (the GCT official classification) to 'High-Speed' status. (again, the official classification). In their Semi-High Speed form they had 7ft 0in wheelbase Brill 21E trucks and 45 hp Westinghouse 323V motors. In High Speed form the wheelbase was 8ft 0in and the motor horsepower was increased to 60, several makes being found throughout the fleet. The Semi-High Speed cars themselves had only undergone a partial modernisation to tide them over before scrapping. The fleet replacement had just started following the entry to service of the Coronation cars but was thwarted by the advent of World War II. The recently upgraded regen cars got a much longer innings than would probably have occurred otherwise but for this war. All of this has become obvious in retrospect!

The production batch of cars was still a vast experiment

The production batch of cars was still a vast experiment to assess propulsive systems for future standardisation. They were, of course, quite the largest fleet of regenerative cars in Britain within this second generation. They were equalled only by the forty-strong fleet with which Birmingham became entangled early in the first generation of regenerative braking, with such disastrous results.

This very large commitment of regenerative cars in Glasgow – while still an enormous experiment – requires a lot of handling and is therefore given a chapter of its own to follow on with.

One of the oldest of Glasgow's 40 production Regen cars was 693, seen here on a football match special to Hampden Park in 1952, shortly before scrapping.
(Photo: STTS collection)

Liverpool 810 on route 22 in the late 1940s. *(Photo: WJ Haynes)*

THE ABORTED LIVERPOOL INTEREST IN REGENERATIVE BRAKING, CAR 810 – 1934

Liverpool Corporation had toyed with the concept of regenerative braking in a desultory fashion, lured on by the reports of massive savings in current consumption elsewhere. Despite being at that time without a Transport General Manager, the City Electrical Engineer, PJ Robinson, was asked to fit regenerative braking into one of the so-called 'Cabin' although also called 'Robinson' class of cars then being constructed at the Edge Lane Works under his overall supervision. This had been all but completed when, on account of some wiring difficulties or snags, it is related, the system was dismantled completely before completion and Car 810 entered

Liverpool never rejoiced in the questionable pleasure of experimentation with regenerative braking during its Second Phase of investigation

service with straight series-wound motors in January 1935.[38] There is the possibility that, as this batch of cars had electrical equipment by Crompton-West, an infringement of an MV. patent may have occurred. If MV had become aware of this they may have threatened proceedings if car 810 ever ran in that form.

So Liverpool never rejoiced in the questionable pleasure of experimentation with regenerative braking during its Second Phase of investigation in British tramcars. The abandonment may have been prompted by the appointment of a new General Manager, WG Marks, who was perhaps less inclined to experimentation (particularly when not started by him).

THE BIRMINGHAM REGENERATIVE BRAKING CAR, 820 OF 1936

With the 1930s revival of tramway interest in regeneration, following the success in Paris, a second chance had been extended to the system. A. W. Maley of Maley & Taunton had the reputation of being an outstanding and very forward looking tramway engineer of that time. He realised that the economic future of street railed transport lay, ultimately, in the ability to harness effectively the regeneration process and decided to approach the matter obliquely. He knew that he was up against a very deeply rooted antagonism towards regeneration. He knew that he would be more than unlikely to get agreement on any such experimentation and sought permission to test what he euphemistically called 'No Resistance Control'. This was another name for 'Field Control' with the realisation that it was but a short step to regenerative braking. It was, in essence, similar to the motoring element of Johnson-Lundell's 1904 scheme, adopted for single-commutator motors. In truth it was 'No SEPARATE Resistance Control' as he masked their existence in the multiple combination field windings he used. This was an idea directly plagiarised from Johnson-Lundell's 1904 scheme! Thus fitted up in 1934, his 'No Resistance Control' went remarkably well and probably only a little gentle persuasion would have been required in explaining that he could safely incorporate regeneration within the established, successful experiment – what with modern techniques, improved accoutrements – not to mention the local trust in a sincere and skilful engineer. He got the 'go-ahead' to try it out on Car 820[40.]

So, with overvoltage and dewirement relays having been carefully incorporated, and involving the air track brakes, Mr Maley added the regenerative element

One of the Maley & Taunton Air Brake Series (812-841) photographed in January 1937 in representation of the likeness of Car 820 which was fitted in 1934 with the Maley & Taunton 'No Resistance Control' or 'Field Control'. AW Maley added regenerative braking in June 1936 in addition to the so-called No-Resistance Control and the car emerged from the Works in September of that year. Motormen experienced much fluctuation of braking performance and the equipment was removed between December 1937 and February 1938.[39]

(Photo: BJ Cross collection)

between June and September 1936 when Car 820 emerged for assessment on the streets of Birmingham. All went extremely well, with no suggestion of any recurrence of past history, for 15 months. However, the motormen had decided that they had experienced too much fluctuation of braking performance in service and the equipment was removed towards the end of December 1937. Car 820 emerged from Kyotts Lake Road Works in February 1938 as, once more, a simple series-wound motored standard electric car with wholly standard equipment.

Number 820 had been one of the 812-841 series of cars built in 1928-29 by Messrs Short Brothers of Rochester and Bedford, on Burnley maximum traction bogies, designed by Arthur C Baker. The braking was of Maley & Taunton manufacture and the underframes supplied by the Brush Electrical Engineering Company. Seating was provided for 62 passengers – 27 in the lower saloon and 35 above. The cars' height was 15ft-10½in and their weight 16 tons 2 quarters. Further information on the Maley & Taunton equipment appears in Appendix 4.

JOHANNESBURG'S EXPERIMENTAL CARS OF 1934 – NUMBERS 2 AND 9

While this book has generally been restricted to the history of regenerative braking as applied to British Tramway concerns, 'Phase Two' of the application of this technology to tramcars would be wholly incomplete without reference to Johannesburg's contribution to this era of British tramway engineering. As interest was developing in Britain, Johannesburg was keeping a very close eye on transport advancement and, indeed, was running neck and neck with Britain in its application.

Two experimental cars were ordered from the Metropolitan Cammell Carriage & Wagon Company, Limited, of Birmingham in 1933 to assess a number of modern applications with a view to the modernisation of that city's ageing tramcar fleet. These included such issues as all-metal bodies: modern four-wheel truck designs: four-wheel trucks versus bogie cars, and regenerative braking versus the traditional hand-cum-rheostatic braking equipments. The cars were numbers 2 and 9 in the Johannesburg Municipal Tramways fleet and were delivered in 1934, 2 as a bogie car and 9 as a single-truck car with EMB hornless, 8ft 6in wheelbase swing-axle design of truck.

There exists dubiety over certain points, mainly as to which car the regenerative equipment was applied. The late Tony Spit – with additional post-mortem material by Brian Patten – gives:

> (of Car 9) 'Electrical equipment supplied by Metropolitan-Vickers – two MV 50 hp motors, Metrovick regenerative remote contactor control'.[Spit., T. & Patton., B.: *Johannesburg Tramways*: 1976: The Light Railway Transport League: pp 92-93]

> (and of Car 2) 'Electric equipment supplied by Metropolitan-Vickers – four MV105, 35 hp motors; Metrovick series remote contactor control' [Ibid: p94]

The Metropolitan-Vickers Electrical Co, Ltd. data sheet 425 for MV116CS motors, does not quote the car number on which they were used, but they were 560Volt motors 'not suitable for operating in permanent series pairs'. It is, however, certain that they were used on car 9.

Taking the M-V. records as probably the more likely to be accurate (while accepting that discrepancies have been known to occur in the best of trade official records) the following statement would seem to be correct:

> Car 2: Double-bogie – Electro-Mechanical Brake Co, Ltd, of West Bromwich, type – Radial Arm, Equal Wheel Bogie, or 'Heavyweight' bogie, nicknamed 'The Jo'burg Truck', with four MV 105 motors and electro-pneumatic remote control. It remained in service until 1959, being sold for scrap in 1960. This was not a regen car.

> Car 9: Single-truck – Electro-Mechanical Brake Co, Ltd, type Swing-Axle 'Hornless', laminated-axle-spring suspended, four-wheel truck of 8ft 6in wheelbase with Skefko spherical roller bearings, two MV116CS 50 hp regen motors and MV remote contactor control.

After a runaway accident in 1950 Car 9 was laid up, only to be scrapped in 1956. This particular truck was finding favour in Britain as a smooth running, readily maintainable four wheeler imparting less noise and vibration, with more comfort, than the standard 21E variety.

Facing page: The strong equatorial sunlight on Johannesburg No.9 casts a shadow on its EMB hornless truck. *(Photo: BT Cooke, courtesy BJ Cross)*

Below left: The Maley & Taunton Swing-Link truck: this example is ex-Liverpool Car 934 (Glasgow 1007) and was photographed in the course of its being scrapped at Coplawhill in 1959. The motors were slung outside the axles, affording, in this case, a fairly short wheelbase of 4ft 2in for the easier negotiation of sharp curves – but at the expense of the disadvantage of greater rotational inertia in the horizontal plane resulting from increased distance of heavy motors from the king-pin fulcrum.
(Photo: Struan J T Robertson)

Below right: The EMB Radial Arm truck: this example is also ex-Liverpool and is also in the process of being scrapped at Coplawhill, still more or less in a complete state. The motors were conventionally slung inside the axles that were carried on the radial arms offering vertical play of the axle-boxes in their horn blocks.
(Photo: Struan JT Robertson)

Although Liverpool was EMB's best customer for their hornless trucks, Edinburgh Corporation very considerately painted the main features in white for this photograph of a sample truck operated by them.
(Photo: STTS collection)

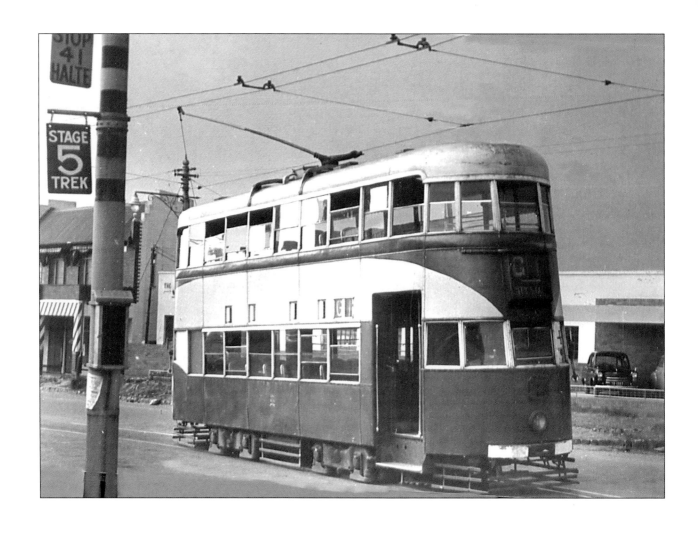

JOHANNESBURG MUNICIPAL TRAMWAYS – 50 REGENERATIVE CARS – 1936

A trait in overall design could, tentatively, have been traced back to the original double-bogie Brush-built Leeds 255 of 1933

In 1935 the Metropolitan Cammell Carriage & Wagon Company of Birmingham received an order from the City of Johannesburg for 50 totally enclosed double-decked, equal-wheel bogie tramcars to be fitted with Metropolitan-Vickers regenerative braking and low-tension electro-pneumatic remote control. They entered service there between July and December 1936 and the cars followed in general design the experimental bogie Car No.2. A trait in overall design could, tentatively, have been traced back to the original double-bogie Brush-built Leeds 255 of 1933. The principle of two motors in pairs, in permanent series, coupled with battery powered electro-pneumatic remote control established a most successful system which politics prevented from widespread perpetration on the streets of Britain. The motors involved were the MV 109DW pattern and the gross dimensions of these substantial vehicles were: 37ft 8in overall length, 7ft 2in wide and 15ft 7in from rail to roof top. The bogies had 27in diameter wheels and seating was provided for 76 passengers – 32 in the lower saloon and 44 upstairs, with an official standing capacity of 16.

The first 26 became 201-226. These were followed by a further 24 filling the vacant fleet numbers of withdrawn single-deck cars and the transfer of single truck cars to native service. There seems to have been no order of allocation of the two types of bogie trucks other than – possibly – the marrying of bodies to bogies in the order of delivery of each. The bodies were of modern, all-metal, lightweight construction, with steel framework and aluminium panelling, featuring partially streamlined ends. There were two types of bogies employed. The first were the Electro-Mechanical Brake Company (EMB) radial arm equal wheel variety already featured in the second series of LCC HR/2 trams and also in the Liverpool system. The others were by Maley & Taunton whose 'Swing Link' design offered the possibility of shortening the bogie wheelbase with the mounting of the motors outwith the axles.

A fault in the wiring of the EMB cars led to the brakes being released instead of being applied when the controller handle was brought back and led to two cars overturning on hilly road runaways, fortunately without fatality. This was soon corrected but the EMB cars seemed to remain prone to runaway accidents even so! Johannesburg's system closed in 1961 but the operators must have been satisfied with their car sets of regenerative equipment as their cars retained this for the whole of their lives. Several cars, indeed, took part in the closure procession of 1961 and one of the regenerative fleet is preserved in the James Hall Museum of Transport in that city.

It has been much observed that there was a similarity of body construction that could be traced from the London HR/2 cars through the Leeds bogie car of 1933 then the Liverpool bogie cars, their Streamliners and ultimately to the Glasgow Coronations. They were all semi-streamlined, all equal-wheel double-bogie cars with four motors paired in permanent series coupling, suggesting a degree of almost family descendency. It would be wrong, however, to suggest any tendency of copying designs. Instead each shows its own personal individualism but the equipment, trucks, and so forth, simply reflect the designer's choice out of a strictly limited pool of availability at the time.

It should not be forgotten that Metropolitan-Vickers also exported MV119DY regenerative motors to Auckland, New Zealand, MV116ES regenerative motors to Oslo and MV 301BZ Field Control motors to Rotterdam, thus highlighting overseas exportation.[42]

Facing page upper: Johannesburg 223 at Malvern terminus. The car is equipped with Maley & Taunton bogies.
(Photo: P Eaton, courtesy BJ Cross)

Facing page lower: Johannesburg 226 at Market Street. This car has EMB Heavyweight bogies.
(Photo: BT Cooke, courtesy BJ Cross)

Loose Ends

There may well be many more regenerative tramcars during the Second Phase of Regeneration in tramcars that simply have not come to light, but it is our contention that the important ones have been covered. However, there are one or two odds and ends that still must be mentioned. Sidney S. Guy's letter of 7th January,1931 has already been referred to in Chapter 4.[43] In correspondence with John Markham regarding further evidence of Regen. experiments in London, he was able to confirm the following:

> "As far as London 1418 is concerned ...I have (somewhere) copies of the M-V illustrated sales leaflet featuring electro-pneumatic equipment on a London E/1 tramcar, which I am pretty sure was regenerative (from the photographs of the rear of the switchgroup), I can only conclude that it was done in the London Transport days post 1st July, 1933. It may even have been the set removed from Glasgow 305, or contemporary with it. I also remember writing to Ted Oakley a while back on this."

The answer to that previous letter to Ted Oakley was then sought and John Markham produced it, as follows:-

> "E/1 car 1418 with MV116DV motors. My list refers to this as MV114DV, but it is a written copy of a copy. So! I suppose that it is possible that it could have been MV116DV.However, I notice on the Motor Data Sheet that it is shown as for LCC Tramways, and dated 29-1-34, but by then London Transport had taken over the lot as from 1-7-33. Interesting!" [44]

Cryptic, and unhelpful! – but almost certainly indicative of another, unrelated London regenerative braking tramcar, in 1418. Then there is a line on another London regenerative car in a Metropolitan-Vickers Traction Bureau Sheet 423 of 11th June 1943. This indicates Motor No.MV114CR as having compound fields but without any running number identity, type, class or pre-LPTB ownership. [45]

Now for two very definite – but amazingly little-heard of regenerative braking cars in Edinburgh:-

Edinburgh, Maley & Taunton Regenerative Braking Cars 81 & 162

A most interesting note has been drawn to attention by Mr Alan W Brotchie which demands quotation in full:-

'Old Standard' Car 81 was fitted, in February 1935 with experimental Maley regenerative system: the wheel-brake valve was replaced by a deep 'D'-shaped valve-box with 2 brake and 5 regenerative notches and 'OFF'. The old air-wheel brake handle was used on it and the box attached to the controller barrel by a curved hollow bracket. Regeneration was applied by movement to the left, and then air braking by movement to the right, passing 'OFF', on the way. This equipment was removed in January 1937 and was refitted in March 1937 but with the air-wheel notch now a further notch in the same direction as the movement for regeneration. Also, a small controller type handle was provided for this valve box. All this equipment was removed from Car 81 which then got normal electrical and brake equipment in January 1938. It was re-fitted in February 1938 to 'Old Standard' Car 162. Car 62 was in service thus equipped in the early part of 1938 and then lay unused until May 1939 when the equipment was removed and normal brake and electrical equipment of 'tapped-field' type was fitted' [46] It is likely that this was but the tip of the iceberg and that alterations to motors and controllers were also carried out.

Around this time, 1934-38, AW Maley, of Messrs. Maley & Taunton, Wednesbury, as we have seen, was investigating the use of heavy motor field coils to improve tramcar power and economy through the elimination of wasteful resistors, to which he gave the name of 'No Resistance Control'. Realising the principle was but a short step from regenerative control, he projected his research into this field and built a couple of experimental regenerative motor equipments, one set of which was tried out – as above – in Edinburgh on Cars 81 and 162, and the other, with clever persuasion, on Birmingham Car 820 already referred to. [47]

(Tapped-fields are further discussed in Appendix 3)

Having dealt with the vehicles of the Second Phase of Regenerative Braking in tramcars in this country for consideration of some of the overt technology of the Phase, specific attention will be drawn to its largest fleet of regenerative cars, that of Glasgow's forty regenerative braking cars. We must not forget, however, the fifty-strong, and very successful Johannesburg fleet, also designed and built in England, as we have seen. The Second Phase's period of progressive activity in the tramway world, together with such spin-offs as field weakening, was essentially confined to the 1930s. The Second World War brought an abrupt cessation to all such theory but, as we shall see, many of the regenerative braking cars soldiered on in service for at least a decade.

They had to. There was nothing else for it during war-time.

REFERENCES – CHAPTER 5

1. Webb, J.S. & Groves, F.P.: City of Birmingham Tramways Co., Ltd., - 2: in Tramway Review: 1992 – Summer: 150: 214-225 (See p222).: Courtesy F. Philip Groves
2. M.T.T.A. Minutes 1932/33: Item XVI – Tramways – Regenerative Control: pp57-58 and 30:courtesy of B.M. Longworth
3. M.T.T.A. Report of Council for 1933-34: Item X – Regenerative Control for Tramcars: p19: courtesy of B.M. Longworth
4. M.T.T.A Minutes 1931-32: Item XII: p.37: Courtesy of B.M. Longworth
5. M.T.T.A. Conference 22/24-6-32: 1931-32: Item XII: Regenerative Control: p.37: Courtesy of B.M. Longworth
6. M.T.T.A. Report 1932-33: Item XVI – Tramways: Regenerative Control: 57-58: Courtesy of B.M. Longworth
7. M.T.T.A. (Managers' Section): 10/11 September 1931: pp3-38 (see p 11): Diagram 4. Fletcher, G.H.: Tramway Regeneration: in The Electric Railway, Bus & Tram Journal: 1931-Sept 18: pp 132-145 (see p.135): Diagram 4
8. Groves., F.P.: personal communication: 23-11-2001
9. M.T.T.A Minutes 1932-33: Item XVI – Tramways Regeneration Control: p.57
10. M.T.T.A. Report of Council for 1932-33: Item X - Regenerative Control of Tramcars: pp 16-19.: (see p.17).
11. Metropolitan-Vickers Electrical Company, Limited, Attercliffe Common Works, Sheffield – Letter of 7-10-32 from R. P. Knight, Traction Motor Engineering Department: to Mr Brooks (Mr Ramsden): Topic – Manchester Regenerative Tramcar. R.P.K./G: p.3. Courtesy of F. Philip Groves
12. Metropolitan-Vickers Electrical Company, Limited: Memorandum of 9 May 1933: Topic – M.T.A. Regenerative Equipment – Report of changes to control system for London service: from W.B.G.C./B.M.V. (W.B.G. Collis): Courtesy of F. Philip Groves
13. Groves., F.P.: personal communication: 30-6-2001
14. M.T.T.A. Report of Council for 1933-34: Item X: Regenerative Control for Tramcars: pp 16-19: (see p.18)
15. Metropolitan-Vickers: Traction Bureau – dc Railway Motor Data: Sheet No.407 of 17-11 1931: Courtesy of John D Markham
16. Markham., J.D.: personal communication: 27-9-2003
17. Markham., J.D.: personal communication: 27-8-2002
18. Markham., J.D.: personal communication: 27-9-2003
19. M.T.T.A. Journal 1934: Regenerative Control of Tramcars: p.37. Brotchie., A.W.: personal communication: 13-9-2001
20. M.T.T.A. Journal 1934: Regenerative Control of Tramcars: p.37: Edinburgh Corporation to convert eleven tramcars to Regenerative Control. M.T.T.A. Report of Council for 1933-34: Item X: Regenerative Control for Tramcars: pp 16-19
21. Young., A.D.: Leeds Trams 1939-1959 – Part 3 – The Great Leap Forward: in Modern Tramway: 1972 – August: pp 264-269: (see pp 266/7)
22. M.T.T.A. Journal 1933-34: Item X: Regenerative Control for Tramways: pp 16-19
23. Ibid
24. Markham., J.D.: personal communication: 28-5-2003
25. Ibid
26. Groves., F.P.: personal communication: 23-11-2001
27. Ibid: Groves., F.P.
28. Oakley., E.R.: personal communication with John D. Markham:7-9-2002
29. Groves., F.P.: personal communication: 15-6-2001
30. Ibid: Oakley., E.R.
31. Groves., F.P.: personal communication: 30-6-2001
32. Ramsden, H.K.: Metro-Vick's Engineer's letter of July 1933: File 929250: courtesy of F.Philip Groves.
33. Ibid.
34. The periodical "Speed" – Spring 1933: The Magazine of Edinburgh Transport Department: courtesy of Alan W. Brotchie
35. Gillham., J.C. & Wiseman., R.J.S.: The Tramways of West Yorkshire: p.62.
36. Gillham., J.C.: A History of Halifax Corporation Tramways: in Tramway Review: 1967-Winter: Vol 7: No. 52: pp 89-95
37. Groves., F.P.: personal communication: 15-6-2001
38. Horne, J.B. & Maund, T.B.: Liverpool Transport: Vol 3: 1931-39: p.98.
39. Lawson., P.W.: Birmingham Corporation Tramways Rolling Stock: Part 15: Modern Tramway: 1971-October: pp330-336: (see p.335) Courtesy Charles C. Hall.
40. Markham., J.D.: personal communication (verbally)
41. Spit., T. & Patton., B.: Johannesburg Tramways: pp 81-102 (see p.94)
42. Markham., J.D.: personal communication 28-8-2002
43. Municipal Tramways & Transport Association Minutes of Committee of January, 1931: Appendix No.7: p 20: Letter of 7th January, 1931 from Mr Sydney S. Guy to Mr C. Owen Silvers – to be put before the Committee of the M.T.T.A.
44. Oakley, C.E.: Letter of 7-9-2002 to J.D Markham: courtesy J.D.M.
45. Markham, J.D.: personal communications: 23-8-02 & 27-8-02
46. Hunter, D.L.G.: personal notes: courtesy Alan W. Brotchie
47. Lawson, P.W.: Birmingham Corporation Tramways Rolling Stock, Part 15, "The Air Brake Cars": in Modern Tramway: 1971-October: pp 334-5

THE SECOND PHASE OF REGENERATIVE BRAKING IN GLASGOW

BRITAIN'S LARGEST FLEET OF THESE TRAMCARS – 1934 - 35

The Glasgow Corporation Transport Department carried out the most comprehensive experiments with regenerative cars of all the British systems. The way of the pioneer is not always an easy one and the undertaking suffered its own trials and tribulations. Struan Robertson explains what these were – much from his own first hand observations.

BACKGROUND INFORMATION:

As has been noted, the meteoric rise of the internal combustion motor during the 1920s was threatening to oust the electric tramways to the point of extinction. Indeed the decade of the 1930s saw the majority of the smaller city tramways and their counterparts serving rural areas being closed down and replaced, mainly, by motorbuses. Trolleybuses joined the battle against the trams from 1925 onwards. The tramways of any standing realised that 'the writing was on the wall' and that a great deal had to be done to their fleets of cars if they were to survive. From 1925 to 1927 Glasgow Corporation, as just one of such systems, set in motion an intensive investigation into the entire gamut of questions about what was to be done to these cars. Following a lot of experimental work, they produced in 1927 the first four experimental, modernised cars, 103, 105, 142 and 159. Each featured some of the different main advantages thought to have been an approach to the answer to the whole situation. Practical assessment of those four cars over the rest of the year decided the line of approach to be followed. This involved the re-equipping and re-tooling of their Coplawhill Car Works and also of their Elderslie Depot and Workshops recently acquired from the Paisley District Tramways, in readiness to cope with the modernisation of almost the whole fleet of tramcars. In 1930 these reconditioned cars were being turned out of the Works at the rate of five per week.

[see: The Tramway & Railway World: 1930 – October 16: p.233]

The best of both the Glasgow and ex-Paisley fleets were the responsibility of Coplawhill for full modernisation. This comprised enclosing the open balconies, re-trucking with modern 8ft 0in wheelbase trucks with 27in diameter wheels and 60 hp high-speed motors and air-braking. Also provided were new upholstered seating to match the best that the contemporary motorbuses could offer. As for the older cars, less capable of such radical treatment, bit-by-bit they were sent to Elderslie Depot workshops for a similar, but less extensive fit-out. Indeed they were seen, then, to be approaching the end of their service lives having run for a hard 30 years or so on short wheelbase trucks and inadequately powered for the very intensive service demanded of them. They were also given enclosed balconies, eventually fitted with modern seating and, in a cascading process, given the best of the 7ft 0in wheelbase trucks with 31¾in wheels and semi-high speed motors displaced from fully modernised cars. Following a trip to London by Councillor Paddy Dollan, a batch of 51 maximum traction cars along the lines of the LCC Class E/1 was bought.

Most of this modernisation was drawing to completion by 1934-35 but the tramwaymen realised that this was not nearly enough. Indeed, if the tramways were to survive, they were required to beat the motorbus opposition soundly in terms of comfort, speed and even luxury. Plans were on the drawing board for just such a lovely tramcar – the 'Coronation' to come.

These Coronations were totally enclosed, having folding platform doors, had luxurious seating and rode on equal wheeled, four-motored, bogies with remote control. Excellent vehicles! Everyone liked them. Meanwhile, and while thought was gently maturing on these forthcoming superb additions to the fleet, the tail-end of the modernisation programme was approaching, involving the creation of the final Semi-High Speed Standard cars.

... if the tramways were to survive, they were required to beat the motorbus opposition soundly in terms of comfort, speed and even luxury

THE SECOND PHASE REGENERATIVE BRAKING GLASGOW FLEET OF FORTY CARS, 1934-35:

As described in the technical chapters, Standard Car 305 was fitted with experimental re-wired MV101DR motors reclassified as MV101CR very early on in the nationwide experiments leading to the Second Phase of regenerative braking for tramcars. The car had acquitted itself sufficiently well for a team of hard-headed Glasgow Councillors in the Tramways Committee to agree to order a fleet of 40 regenerative braking cars. Quite obviously, with hindsight, this was aimed at being ultimately applied to the still drawing-board 'Coronations', but it was not to be.

The best of the last pre-modernised Standard cars were taken, their bodies fully modernised and fitted with brand-new 8ft 0in wheelbase, 27in wheel diameter trucks. These were equipped with new British Thomson-Houston Type 101J state-

Car 809 stood for her official photograph on Albert Drive Bridge in pouring rain, as just converted to regenerative braking control, on 17th March 1935. There was no external detail to indicate the change of control mechanism. The destination information was, of course, incorrect, as the car's approach to Springburn would be by way of Hope Street, Cowcaddens Street and Garscube Road rather than via Castle Street. 809 is immediately ex-works, and the shunter would not know his geography. *(Photo: GCT)*

of-the-art, regenerative motors and OK45B controllers. The 40 cars represented the largest fleet of regenerative braking tramcars in Britain at the time. They were placed in service between 15th December 1934, when the first car (807) arrived, and 3rd December 1935 when the last (Car 734) took to the road. The balance of cars still with semi-high speed motors were simply run into the ground. A good number had been scrapped before the Second World War and the survivors had to be kept going for extended war service in the city. The war-time veto on scrapping vehicles throughout the Transport World had clamped down.

[see: Young, A.D.: Modern Tramway: 1972 –October: p348]

Coplawhill did a magnificent job in reconditioning these old, work-weary cars to a modernised regenerative status. Built, as they were between January 1899 and July 1901, they ranged from 33 to 35 years old. They looked good with all traces of body-sag and drooping platforms removed. The frames were straightened up and the coachwork repainted in the full glory of Glasgow's attractive, fully lined-out livery. All carried the Prussian Blue route colour livery.

Naturally, at first, these cars were tried out on several of the Glasgow routes to assess their adaptability. All functioned well enough at first but, as was common to all regenerative cars to date, they began to show little idiosyncracies after some eighteen months or so, and their special equipment was not extended to further production.

... they began to show little idiosyncracies after some eighteen months or so

The 40 vehicles were all blue cars and had originally been allocated to Newlands, Renfrew and Maryhill Depots together with one each from Dennistoun and Dalmarnock. They retained their blue route colour as they were destined to operate from Govan and Possilpark Depots to work the blue, latterly No. 4 service complex, from Possilpark and Springburn, ultimately to Renfrew via Govan. None of the group sustained any war damage although five were involved in collision damage during the excessively heavy 'pea-soup' fog of Wednesday 19th November, 1941. These were 694, 734, 804, 817 and 863.

Glasgow from the mid-1930s had been assessing mercury-arc rectifiers in some of the substations which, while very expensive in capital outlay, had the very great merit of not requiring a constant overseeing engineer as they were capable of automatic operation, once fired. However, it was soon found out that there was an intolerance of the inverted currents of regeneration as the mercury-arc could not function in the reverse direction because the arc became extinguished. When the cars were tried out on the remote south-side of Glasgow working into the Spiersbridge substation area, much mischief occurred. This does add some credence to the view that these cars might actually have been intended to operate the lengthy (also blue) service from Renfrew Ferry via Barrhead to Milngavie that needed around 40 cars. It had lots of gradients to negotiate and was the only all-day service running through the city relying on Semi-High Speed cars. There was the potential to hold up following high-speed cars, so this service was ripe for equipping with faster rolling stock.

As will be seen in Chapter 13, in the late 1930s further trouble arose with the regen cars with the welding of the controller contacts. Despite the attention and recommendations from Metropolitan-Vickers, any outcome has been lost owing to the impending outbreak of World War II. Whatever was done, the cars continued to give trouble until in the post-war years the regenerative shunt-fields, resistors and so forth were removed after some 12-15 years service and most cars continued life for a short while as 'Regen Convert' motor machines until scrapped.

The majority remained as regenerative cars for up to 15 years. One car, 747, lasted 16 years in its regenerative state. The list of longevities is as follows:

Cars retaining regen. equipment for		
12 years	3	(converted in 1947)
13 years	6	(converted in 1948)
14 years	7	(converted in 1949)
15 years	13	(converted in 1950)
16 years	1	(converted in 1951)

Cars recorded as regen-conv but no date recorded	5
Cars never recorded as reconverted to standard	4
Plus Car 698 Converted to Field Control equipment	1
Total	40

THE MAJOR REGENERATIVE BRAKING EXERCISE IN GLASGOW:

The advent of the fleet of regenerative cars in Glasgow was heralded by the arrival of Car 807 when it was turned out of the Coplawhill Works on 15th December 1934. Naturally, the car was given immediately to the engineering staff for investigation and familiarisation before handing on to the Driver Training School for the commencement of outdoor staff training. It would be some time before the regenerative exercise got under way. January 1935 saw the next two cars emerge, 830 and 859, that were probably also annexed by the Driver Training school for very necessary training of already expert motormen in the handling of this entirely new and different style of controlling the cars and the learning of new techniques. It was not until February 1935 that Coplawhill really began to get into its stride, for by then the suppliers of the equipment were delivering batches of new motors, controllers and gear. Eight cars emerged in that month while the peak of output was achieved in March and April with eleven and twelve cars, respectively. The main fleet had been rounded off in May with three more while the three last stragglers appeared in August, September and December.

The bulk of the regenerative fleet therefore appeared during the period February to April of 1935 by which time a team of special motormen had been trained and the regenerative 'exercise' got under way, revenue earning. Breaking-in troubles were inevitable as the cohort of specially trained motormen grew accustomed to their radically different charges. Sadly, it would soon have become apparent, first to the Engineering, and then to the Tramways Committee, that, despite all the definite advantages that regeneration offered, the system was yet too immature for extensive incorporation in any projected large fleet of ultra-modern construction. Amongst the first irritations to be faced, possibly, was that of electric light bulb failures. Unaccustomed to the very high surges of overvoltage from regeneration, the standard tramway rough service light bulbs of 95 Volts and 40 Watts in series of six (on the same principle of Christmas-tree lighting) gave up the ghost before recurrent peaks of overvoltage requiring lads with boxes of spare bulbs to be stationed at strategic points, especially at the foot of Hope Street. They would hop on to the cars and replace any failed bulbs. One can envisage the catastrophic effect on time-keeping that such would have precipitated. Probably the spare-bulbs lad would have had to travel with each car, re-fitting in transit, and then getting the next

Standard Car 706 at Springburn Terminus around 1947-48. This car was fitted with regenerative braking equipment in February 1935 and, while no date of re-conversion to simple series fields has been ascertained, this would – almost certainly – have been during the intensive period of such treatment between 1948 and 1950. During this time there were 26 known conversions representing 65% of the total "Regen" cars undergoing such modification. It is almost certain, then, that car 706 was a regenerative-brake fitted car at the time of the photograph. The photograph was taken at the east end of Hawthorn Street opposite the then "Boundary Bar" – now long gone – with the Balgrayhill "genteel" houses in the background. The car is showing the No. 27 service number. This was introduced in 1943 from Springburn to Shieldhall or Renfrew Cross, essentially a subdivision of the Service 4 complex covering generally the same main course but deviating at its north end to the Hawthorn Street terminus.

(Photo: BJ Cross collection)

available car back to base for the ensuing victims! All of this must have represented an added on-cost of unforeseen, but formidable extent! However, the passengers, generally, expressed their appreciation of the increased speed of point-to-point transit, the result of the rapid acceleration of regeneration. Yet – as especially in Edinburgh – they pointedly objected to the jerkiness of step-down voltage braking, again inherent in the early days of regenerative braking.

Worse, as we have seen, was to come! Glasgow's new and expensive mercury-arc rectifiers were experiencing shut-down troubles. The voltage generated by the regen cars exceeded that of the substation and attempted to reverse the current flow. This damaged the substation machinery. Even in the older rotary-converter substations, this very high back-pressure, especially from a number of regenerating cars, could disrupt the converters and so back along the 3-phase delivery cables to the primary plant in the generating station. Just as in Edinburgh, so in Glasgow the regenerative fleets became confined to a route (or routes) which were totally supplied by rotary converters.

The voltage generated by the regen cars exceeded that of the substation and attempted to reverse the current flow

Two Blue Standard cars on service 4 crossing George V Bridge around June 1938. The leading car is 688 and is a Regen example.
(Photo: DC Thomson, courtesy AW Brotchie)

COPLAWHILL TERMINOLOGY:

At this point it becomes opportune to look at the complexities of Coplawhill parlance! We have the term "Regen Conversion" which referred to the 1934-35 conversion from standard to regenerative braking. On the other hand 'Regen Converted' – so very similar, and yet so totally different – indicated the much later process of reconfiguration of the compound field windings to produce a simple, series-wound machine, virtually identical to the very useful MV101 series of motors from which the 101J had been designed. Up till the 1920s or so, there was very little, if any, swapping around of trucks. A car body and its truck was a marriage looked upon devoutly not to be broken. This resulted in a car body lying around unused in Coplawhill awaiting completion of major repairs or alteration to its personal truck, instead of sallying forth immediately to revenue earning upon a different, fully reconditioned truck. It was only when the early electrified horse cars were being scrapped that a surplus of trucks became available for re-use and the interchange of equipment only arose again after the 1930s modernisation. By that time large numbers of new 8ft 0in wheelbase trucks were being bought in whole for fitting to modernised Standard cars. New trucks were given the numbers of the old trucks they replaced. With the re-trucking of the 'regen' car bodies, 40 old 7ft 0in wheelbase trucks that came off them became redundant and started a re-shuffle around the remaining 'Semi-High Speed' cars.

THE ROUTES & SERVICES OF THE GLASGOW 'REGEN' TRAMCARS:

The 'Regen' cars – apart from a few haphazard cruises into other areas such as the No. 2 Provanmill-Polmadie service and the Giffnock-Spiersbridge portion of the system – were all allocated to the working of the Blue No. 4 service out of Possilpark and Govan Depots. The No. 4 service consisted of a group of intermingled workings, mostly using the main common stem of the Round Toll – Garscube Road – Cowcaddens – Hope Street – George V Bridge – Nelson Street – Paisley Road Toll – Govan Road, and thence westwards to Linthouse, Shieldhall or Renfrew South (formerly Renfrew 'Aerodrome').

Service 4 commenced in Keppochhill Road, in the north, and terminated at Renfrew South at the old Renfrew Aerodrome terminus. The 4A service commenced at Springburn and terminated at Linthouse while the 4B commenced at Lambhill also terminating at Linthouse. There were other modifications from time to time, most

Standard car 741 is at the post-war Renfrew South terminus laying over after a long run from Springburn.
(Photo: FNT Lloyd-Jones/ OnLine Transport Archive)

using the trans-city stem. In 1943 the 4B became the 31 and the 4A, service 27.

The regenerative cars had all originally been blue cars and were culled from various depots as has been seen. Following conversion they were redistributed evenly between Possilpark and Govan Depots where they retained their blue route colours up until the advent of 'Standard Green' (or 'Bus Green') generally: a process started before World War II but not completed until 1952, gradually involving the whole fleet.

The route eventually chosen for the Regenerative Braking experiment is outlined in heavier lines on the accompanying map. It was chosen carefully to test the cars rigorously over a complete cross-section of the Glasgow terrain while also utilising a route largely serviced by rotary converters to assess the cars' capability from all aspects and under all conditions and circumstances.

The portion south of the Clyde and to the west of the city lay largely upon the alluvial plain between the rivers Cart and Clyde. Most of this area is absolutely flat and, in the 1930s, fairly sparsely populated between Renfrew, in the west, and Linthouse and Govan. This kind of terrain allowed high speeds to be sustained in urban service with relatively few stopping places.

Immediately the Clyde was crossed, northbound, a totally different terrain presented itself. Composed of post-glacial drift deposition into drumlins and hills of quite steep gradients, here the city centre was traversed in an almost constant climb through the business centre to the residential north. The route climbed in successive steps and severe inclines to its northern termini. It made the perfect testing ground for the fleet of 'Regen' cars to show their paces, traversing high density shopping and business populations of the intensive commuter demand centrally, to the equally high residential populations peripherally with only morning and evening high travel demand.

Such would have tried out the 'Regen' concept to its fullest extent and proffered valuable assessment of its capabilities and weak points. It did not seem unlikely that such a magnificent experiment was undertaken without a long-term view towards the possible ultimate conversion of the whole fleet in the then not too distant future! The war was not appreciably imminent, just then.

Not only did the route offer fairly long-distance, uninterrupted running on the South Side, for a tramway, to gauge the economy of fast running without the 'surge-

The Blue route is shown by the heavier line on the Route Diagram reproduced below

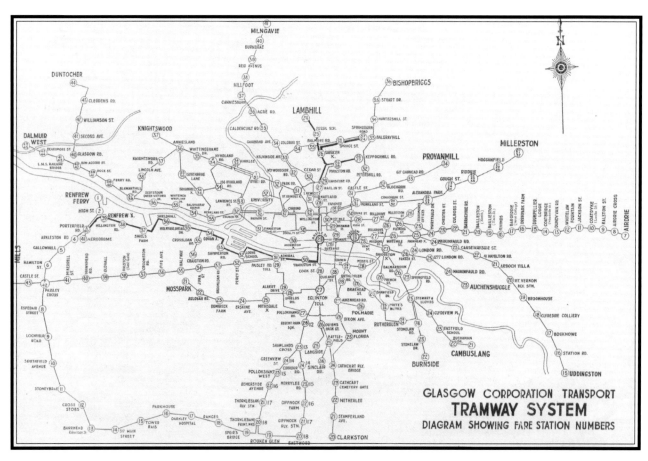

GLASGOW CORPORATION TRANSPORT
TRAMWAY SYSTEM
DIAGRAM SHOWING FARE STATION NUMBERS

Here was an excellent opportunity to exploit the full range of the eight economical controller notches

and-coast' principles of standard motoring, it also offered the 'inching-and-braking' of city centre tradition – the most expensive usage of traction current on any system. It was able to harness the economy of regeneration on the long slopes down from the hills to the north of the city centre, coupled with the overall high degree of car control at moderate, instead of high, cost. Here was an excellent opportunity to exploit the full range of the eight economical controller notches as opposed to only the two on the controllers of the standard cars!

THE 'REGEN CONVERTED' MOTORS OF THE GLASGOW PERMANENT WAY DEPARTMENT:

The earlier Permanent Way Department (PWD) cars numbered up to No.13 were mostly scrapped during the 1930s purge terminating in the advent of the Second World War. There were exceptions, of course, such as No.1-No.3 and No.6 of which No.1 and No.3 have been preserved. Of the rest, due to the post-World War II re-conversion of the regenerative braking equipments of the regen fleet, the majority of the PWD fleet was able to be re-equipped with their modified and redundant BT-H OK45B controllers and 101J motors when the donor trams were withdrawn. In this condition they were virtually identical to the standard 60 hp MV101DR motors in general use and they could keep pace with the passenger cars in service when they occasionally ventured out in daylight. This avoided causing hold-ups through the sluggish function of their old motors. Their lifespan in converted form was sadly short following rejuvenation, amounting to around ten years, but it was worthwhile!

A BT-H Type OK45B controller, still with EMB air-brake interlock valve box fitted, but devoid of air-brake pipes. This is Glasgow Permanent Way Department Tool Van No.27 (No. 2 end) while in the Coplawhill Car Works on 20th September, 1960. *(Photo: Struan J T Robertson)*

THE 'REGEN-CONVERTED' BT-H MOTORS OF GLASGOW'S PERMANENT WAY DEPARTMENT:

The Author's Observations of the Regenerative Braking Cars:

Note: "B" = Old Blue. "SB" = unlined, or Standard Blue. "SG" = Standard, or "Bus" Green.
e.g. Car 688, below, Old Blue, Service 4: Standard/Bus Green, Service 1, (July 1941)

Car No.	Colour & Route Observations
688	B4: SG1 (7/41)
689	B-: SG4 (7/41):31 (9/44):4 (10/44)
690	SG31 (9/44)
692	SB4 (10/40): SG27 (9/44)
693	B4: SB4 (11/44)
694	SG4 (9/44): do. (11/44)
698	B6 (8/44): SB10A (11/44): 6 (11/44) (31/10/49)
	The 2 doors on the side of the dash noticed (9/49)
703	SB25 (11/44): 16 (11/44): 33 (3/49)
706	SG4 (8/44): do. (9/44): 31 (10/44): 4 (1/46)
714	SB4 (8/44). (10/44). (11/44). (9/51)
715	G4
722	B4. SB4 (11/41). (9/44): 4 (5/50) > PWD40
724	Not seen
734	SB4 (11/44)
741	B-. SG25 (9/45)
747	G4.
757	SB4 (4/45)
775	SG4 (12/44)
777	B4A. SG31 (10/44): 27 (11/44)
787	SB27 (8/44). (3/50)
804	B. SB25 (5/49)
807	B. SB16 (8/44): 6 (8/44): 4 (8/44)
809	B4. SB27 (8/44)
811	SB4 (9/44)
814	SB4 (8/44). (10/44). > PWD23
816	SB4 (8/44) (11/44)
817	B4. SB4 (10/44)
823	G4. SG19 (10/44). (11/44)
829	G4.
830	SG4 (11/44): 33 (2/49)
841	G4. SG8 (11/44)
845	B4. SB4 (10/40). (8/44). (10/44)
859	B4.
860	B4. SB4 (8/44)
863	SG-. (8/44)
871	SB19 (10/44)
873	SG4 (9/44): 14 (9/44): 31 (10/44)
875	B4: 16: 4. SB4 (10/44). (11/44)
877	SB4 (7/45). (9/45)
305	SG32 (8/44). (9/44): 33 (9/45)

The author's personal interest in recording the cars on the road stemmed in earnest from the Spring of 1939, just as a lot of historical detail began to disappear under the influence of the post-modernisation scrapping programme. This had really commenced in 1938 with the withdrawal of 92 – the last passenger electrified horse car in service in 1933 on the Abbotsinch shuttle. Most of the cars in those days were still in the delightful old, fully-lined-out livery, and the services enumeration, for a long time crazily inappropriate, had been overhauled and simplified. It was not, however, until the early 1941 period that the dating of these personal observations was regularly undertaken in an attempt to tally potential war losses. Apart from relatively minor damage, and the reclamation of stranded cars from the Duntocher route by de-trucking them and delivering them to the Car Works on motor lorries, and the towing away of similarly stranded 'Coronation' cars at Dalmuir terminus following the Clydebank 'Blitz' there had only been one solitary total write-off during the war (Car 6).

Following the removal of their 'regen' equipment, the great majority being completed over the period 1948-51, the cars continued more or less until their end on the same services, at least working from the same depots, except for incidental emergency requiring spare cars to fill-in elsewhere. By this time they were capable of ubiquitous service. Two of them were cut down to single-deck status for work as Tool Vans as the older tool-vans began to deteriorate. Their windows were simply partly painted over with the Departmental red-brown as was the whole of their bodies.

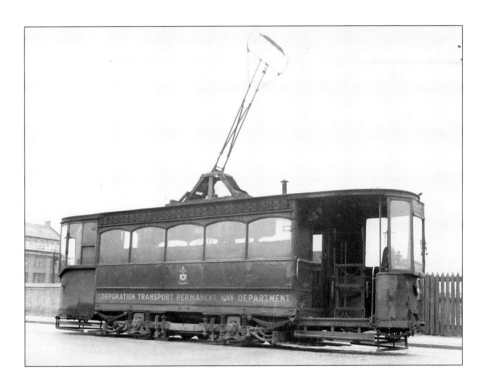

Regen car 814 converted to Permanent Way Department Tool Car No.23. It was withdrawn from passenger traffic on 9th March, 1952 and converted under Order 35/51 of Coplawhill Car Works, between 3rd and 10th November of that year. The car was scrapped at the Car Works on 21st September, 1962.

Permanent Way Department Tool Van No.40, ex-Regen car 722 was converted between 30th August and 4th November, 1954 at Coplawhill Car Works. It was sold to Hector MacKenzie on 28th December, 1961 and despatched to Newtonhill as a source of spare parts for preservation schemes.

(Both photographs are by Struan Robertson)

119

THE ATTRIBUTES AND DRAWBACKS OF REGENERATIVE BRAKING

The technology involved in regenerative braking seemed to be the solution to so many problems but, ultimately, in its first and second phases, was not up to the job. Struan Robertson explains.

A way back in September 1908 the Tramways & Light Railways Association (T&LRA) drew up a remarkable report on the entire field of tramway braking in Britain. Lt Col Yorke, in his Report to the Board of Trade on the Highgate Archway tramway accident on 23rd June, 1906, had suggested the desirability for the Tramway Associations to consider in depth the whole question of tramway braking apparatus and the sanding arrangements for tramcars. This accident to Metropolitan Electric Tramways double-deck bogie car No.115 had occurred on a straight road of decreasing downhill gradient from 1 in 18 to 1 in 27 inclination. It had been the result of an unskilled motorman incorrectly manipulating the hand and magnetic brakes to wear flats on the skidding wheels which then left them locked and, with the reversing key in the wrong position, he jumped off!

The ultimate 1908 T&LRA Report ran to 127 pages. It contained fascinating descriptions of the entire field under review, from service stops through emergency stops and run-away stops, and the available means of braking them. In addition there were tabulated reports on the major tramway accidents to that date and the intense research and testing done by various members of its Committee. The description of the Phase One (Raworth) Regenerative Braking Control makes most interesting reading:-

'The retardation of this type of brake is obtained by the use of shunt-wound motors made to act as generators when descending gradients, or when stopping the car.

Regenerative control covers a good deal more than braking and cannot be judged from that aspect alone. It depends upon the use of shunt-wound motors, which tend to run at constant speed. Weakening the field increases the speed, and *vice versa*, without much affecting the work done so that a car equipped with such motors will run with varying loads and on varying gradients at a nearly even speed for any particular field adjustment. On a falling gradient the car will accelerate until the emf of the motor exceeds that of the line by a sufficient amount to produce a generated current absorbing the work done by gravity. The current so generated is delivered to the line and may be utilised in driving other cars. This is a saving of power and reduces the heating effect, but does not directly affect the braking results which are similar to those of the rheostatic brake. There are some important advantageous differences:

First, the regenerative effect comes into action automatically upon a small increase of speed occurring and is proportional to the excess of speed above that corresponding to the controller position, every controller notch corresponding to some pre-arranged normal speed. Secondly, the motor connections are not allowed to change from driving to braking, the field has not to be built-up before the braking commences, and is strengthened or weakened by the controller independently of the braking current. Thirdly, the electrical circuit is exactly the same for both driving and braking actions, is therefore always under test and any defect makes itself apparent at once.

The main disadvantage is that the regenerative action and its braking effect cease at some minimum speed, which is in ordinary cases about 4 mph. A series field winding may be provided for use at lower speeds, or at any speed, should the line connection fail.

This is simply adding the rheostatic to the regenerative brake. The limitations of braking power by the wheel adhesion and the liability to paralysis by skidding of wheels by other brakes are exactly as with the rheostatic brake. The general question of series versus shunt motors for tramway use involves much more than the braking question.

The greatest advantage of regenerative braking is that it limits the speed on falling gradients by automatic means, provided that gravity acceleration is not in excess of the wheel adhesion at the speed at which the gradient is worked. If the wheel adhesion is too low the wheels will not be skidded fast, but will skid and roll alternately with less injurious results than if they were skidded. The heating of the motors will be comparable to that produced by climbing the same gradient at the same speed. The total heating or final temperature on a given route and service as compared with series motors depends largely upon the driving current. Practical experience shows that on routes with moderate gradients regenerative motors run no warmer than similar series motors, but where there are long and heavy gradients, they run hotter. There are other considerations outside braking, affecting the choice, but as far as that goes the speed control on down gradients is a very valuable feature of the system, especially where the gradients are distinguished rather by their length than by their severity. As the effect ceases at a fairly high minimum speed, regenerative control cannot by itself be used for service stops, nor for emergency traffic stops.

The only form of regenerative control in actual operation in this country is that known as 'Raworth's'. The motors in this system are arranged to work on the series parallel system – that is to say, at the lower speeds the motors are kept in series, while at the higher speeds they are in parallel.

The shunt-winding is supplied with current direct from the line, and in this respect the motors may be said to be separately excited.

When the motors are in series they are worked as shunt-wound motors (separately excited), and when in parallel a few series coils are brought into use, thereby making the motors compound-wound for the higher speeds.

If the speed of a car when descending a gradient exceeds that proper to the controller notch in use, the emf rises, energy will be returned to the line proportionately to the slope, and the speed will remain constant at the highest point. In like manner, if the controller handle is moved backwards, the field strength will be increased and the car will slow-up until its speed is reduced to a speed corresponding to the position of the controller handle.

To return current to the line with regenerative control is absolutely dependent upon the line circuit being unbroken and upon revolution of the car wheels. This makes a second brake indispensable for bringing the car to rest and holding it on a gradient.

An emergency brake independent of the line is provided by introducing the series-winding and short-circuiting the motors through a resistance.

ADVANTAGES:

(a)   Simplicity of adjustment
(b)   Delicate graduation of application
(c)   Utilises energy of the car
(d)   Braking effect obtained from armature action
(e)   Power returned to line

DISADVANTAGES:

(a)   Inoperative at low speeds
(b)   High initial cost
(c)   High cost of maintenance
(d)   Slight tendency to skid wheels
(e)   Braking effect dependent on revolution of wheels
(f)   Tendency to increase temperature of motors and consequent deterioration
(g)   Large number of contact points in electrical circuit
(h)   Dependent on continuity of line current

REMARKS

A perfect system of regenerative control would supply a most efficient form of speed control. As long as the electric circuit is maintained, and the gradients are within the limits of working (ie adhesion) the fixed maximum speed of the car cannot be exceeded. In braking, the kinetic energy, instead of being wasted, is converted into electrical energy and returned to the line.

The number of cars fitted with regenerative control is small, and the Committee has not sufficient experience of their working to make any recommendations in regard to the general efficiency of the system. There is no doubt, however, that a quick-acting brake independent of the electrical equipment should be provided on cars fitted with this type of control.

The Committee is of the opinion that the merits of this form of control are such that it deserves the careful consideration of those interested in electric traction'.[1]

This important Report covered many of the Association's Member Systems, including also many of JS Raworth's Phase One regenerative cars. These cars came under a great deal of conflicting assessment, due not to the system concerned – so very markedly before their time – but to failures of the primitive motors due to their need for conversion from series-wound to compound status. Their inadequate ventilation led to a great deal of overheating and the lack of interpoles (or commutating poles) led to much sparking of commutation from the high regenerative currents. Finally there was the unrelenting jarring vibration in negotiating the sharp turn-outs of points and corners that was thought to have been a contribution to a number of regenerative car accidents. John Markham outlines an interesting and very apt theory of the rupture of field-winding connections. This was due to a combination of overheating of the motors and the acute jerking of cars negotiating very sharp radius bends, especially on single lines with recurrent passing loops, simply fracturing the field connections due to – perhaps – a combination of these. Should this have occurred on a downhill grade the regenerative circuit would immediately have been broken and would render those brakes totally ineffective. The car would inevitably have run away. Certainly Rawtenstall's accident was due to such a cause.

Birmingham's fleet of 40 Phase One regenerative equipped cars on the Raworth principle did not last for many years. Webb, in his book of 1974 blames this on overvoltage 'blowing' of the cars' circuit-breakers, and the dramatic dimming of lights, and serious loss of speed of these vehicles. He reckoned that their modification to standard took place in the 1908-1909 period.[2] Groves, on the other hand, states that Birmingham Corporation prohibited the use of these Company cars on their tracks from early in 1907, mainly due to the casualties caused to the lamp bulbs. [3]

The Raworth scheme did not allow the motorman to throw OFF from maximum speed without first using the regen brake to slow the car down to the running speed of the first parallel notch. Raworth used the motors as pure shunt machines when in the series combination. Throwing OFF at full speed would connect the motors in

City of Birmingham car 209 on the Bristol Road.
(Photo: BJ Cross collection)

series, but each would remain generating at or near full line voltage at the time when the controller was passing over the series notches. Motormen would instinctively throw OFF at whatever speed they were running should an unexpected emergency arise. Thus overvoltages on Raworth cars were inevitable. The Johnson-Lundell system had a distinct advantage here in that motormen could throw OFF without using regen at all.

The second generation of the regenerative system was designed to incorporate a full range of magnetic braking equal to the best standard equipments, in a combination allowing for normal stopping, together with speed control in traffic by the regenerative brake with its recognised marked economy in energy. For rapid stops the magnetic brake was immediately at disposal. Regenerative braking is a self-eliminating procedure, tailing off notch by notch, and therefore strength by strength before fading out altogether around 4 to 5 mph whereupon the airbrake would have been used to bring the car to a standstill, typically, later in Glasgow. Theoretically this was supposed to be followed by the hand-brake to keep the car stationary, but was generally only applied at termini. Elsewhere, the magnetic brake (rheostatic) could equally have fulfilled this function. In practice regeneration was usually terminated at from 15 to 12 mph for competency.

Despite all this, the Association's Report gave a most balanced description of the attributes and drawbacks of regenerative braking in 1908. These remained applicable to the Second Phase examples of the 1930s except for the provision of more modern motors with more sophisticated regenerative equipment, unavailable in those early days.

PROGRESS – BUT STILL SHORT OF PERFECTION – THE 1930S:

The tramcar, as a railed street vehicle, is as much if not even more obstructed by the great predominance of the internal combustion street vehicle, crossing and criss-crossing relentlessly, in search of an opening for overtaking. The requirement of instant braking and rapid re-acceleration is therefore paramount both to avoid collision and to maintain scheduled service speed. Economy of current usage, always desirable, became – in post war slump conditions – a necessity, and the ratio of the complement of passengers to road-space occupation led to political quandary and the throwing in of the towel in terms of the de-restriction of the motor bus.

It was extremely sad that the regenerative braking experiments of the second quarter of the 20th Century collapsed through, even then, immaturity of design of the regenerative motor and its control. The evolution of electronics was still writhing *in utero* of electric concept. Indeed, the thermionic valve still ruled the striving world of electronics and it was not until it started to mature into solid-state electronics in the terminal quarter of the 20th Century that regenerative braking of electric tramcars commenced in earnest to catch up with its usage by "Big Brother" – the world's railways.

Speed being basic in the promulgation of most transport, and energy being a function of the square of the speed, this rendered the least trivial increase in speed, following braking in traffic, to command a vastly disproportionate increase in energy consumption. This had discouraged the development of faster series motors for tramway usage. The greater the energy put into acceleration, the more was squandered on subsequent braking. The obvious answer stared one in the face – regenerative braking! Every engineer knew that, but government, in the shape of the 1911-1912 Board of Trade, had vetoed its use. The result for most tramways in the face of the soaring rise of the much cheaper internal combustion opposition had been dissolution! The government veto simply had to go by the board. Even government saw this and turned a blind eye to its infringement!

'The foundation principle of all dc regenerative schemes lies in the fact that if the field winding of a dc machine is supplied with suitably controlled current it will act as a generator when driven mechanically. The difference between various systems which have been developed consists mainly in alternative methods of exciting the fields'.[4]

In railway practice their series-wound motors, of that era, were separately excited for regeneration through a line-driven motor-generator set. Alternatively, the use of compound-wound motors with shunt fields, line-fed through a control resistor was universally tramway acceptable. Here the braking is smooth, powerful and instantly applicable as the regenerative circuit is identical to the power circuit and required

'The greater the energy put into acceleration, the more was squandered on subsequent braking. The obvious answer stared one in the face – regenerative braking!'

'The government veto simply had to go by the board. Even government saw this and turned a blind eye to its infringement!'

no circuitry changes. There was therefore no chance of open circuit causing loss of brake power – except through dewirement, derailment or extraneous power failure. Also, the immediate backwards movement of the controller handle applied the braking effort without any switching-off, and enabled the re-establishment of a separate circuit with the inevitable lag in the building up of flux and current before brake application.

The amazing advantage of this is seen in the rapidity and ease of speed control of the car to suit all conditions of traffic, emergency or otherwise. This was achieved with the minimum of motorman effort and the maximum of economy, as no other brakes were necessary and no energy was taken from the line. Instead, every such application of the regenerative brakes fed current back into it instead. Neither mechanical braking, and the inevitable wheel-rail and brake-block attrition, nor current dissipation in resistors, obtruded. The résumé that follows comparing regenerative controllers of the time and subsequently with the standard controllers of the mid-20th Century has been culled from the same Metro-Vick article previously quoted:

Operation	Standard Equipment	Regenerative Equipment
1. Applying power	Notch-up from 'off' position	Notch up from 'off' position
2. Running	On full series, or full parallel notch. Two economical notches	On notches 3, 4 or 5, or notches 8,9,10 & 11 parallel - seven economical notches
3. Slow-down in traffic	Switch-off, apply brakes, notch-up again and repeat frequently	Move-back one or two notches only, and hold required speed
4. Service braking to a stop	Switch-off and apply hand brake	Work back notch-by-notch to minimum regenerative speed, finally switching-off and applying hand brake
5. Emergency stop	Switch-off and apply rheostatic, or magnetic brake	Switch-off and apply rheostatic, or magnetic brake

The first two operations highlighted the great advantage and economy of regenerative equipment, obviously, and had shown through much testing that an energy saving of around 30% could be obtained in service conditions, and the equipment was by no means complicated in comparison with the standard variety. The regenerative equipment incorporated all the advantages and safety factors of standard equipment including the full range of rheostatic, or magnetic brake notches, and its additional equipment consisted only of shunt fields and regulating resistors, the shunt-field contactors for the regulating resistance, and the small operational control drum and fingers for them in the controllers.

Increasing the capacity of the equipment with the introduction of higher-speed motors did, of course, achieve improvement. This was, however, at the expense always of higher power costs arising from the greater currents to be handled. In turn, these would require bigger and heavier hand-operated controllers. Better performance and improved economy became the demand and the answer to this was regeneration.

THE FAILURE OF THE SECOND GENERATION OF REGENERATIVE BRAKING IN TRAMCARS:

The sole aim of the First Phases of regenerative braking was to economise mainly in power consumption. There was no opposition in those days. The expectation of the Second Phase, however, was the hope of the remaining Tramway Movement, in the face of the cataclysmic collapse of the smaller, and rural tramway concerns, to 'keep their heads above water' and remain in viable business. The internal combustion opposition was free of railed restriction of movement and, largely because of this, had virtual monopoly of the streets and roads. It also benefited from the cheapness of purchase and maintenance.

From the electrical aspect, the trolleybus – essentially a tramcar broken loose from its restrictive railed track – had gained great popularity in cities, and had carved a niche for itself in certain rural areas. It was still yoked to its overhead, however, much as the tramcar was to its rails and so eventually succumbed in the early 1970s. The internal combustion motor had gained supremacy! Tram rails were ripped up out of the streets but there was no improvement in public service due to the over-

'The sole aim of the First Phases . . . was to economise mainly in power consumption.'

'The expectation of the Second Phase, however, was the hope of the remaining Tramway Movement, in the face of the cataclysmic collapse of the smaller, and rural tramway concerns, to 'keep their heads above water' and remain in viable business.'

124

congestion of city streets with motor car and bus traffic. It was Stuart Pilcher who had proven in 1930 that 'there is no appreciable difference between the speed of omnibuses, tramcars and trolleybuses'.[5]

The sad failure of both the original Phase of regenerative braking in tramcars and of the Second Phase of the 1930s was undoubtedly due to the fact that contemporary electric technique still fell short of the perfection the system needed, highlighting the amazing technological advancement of both John Raworth and, despite commercial failure to make the grade, of the Johnson-Lundell firm in producing working models a century ago. The system required today's solid-state electronics for it really to thrive and produce the wonderful vehicle of the contemporary generation that can be seen on the streets of the continent and in the few lucky UK cities, handling bulk street passenger demand, with low loading facilities and greatest comfort.

THE INCOMPATIBILITY OF REGENERATIVE BRAKING WITH MERCURY-ARC RECTIFIERS:

And so we arrive at Mercury-arc Rectifiers (MaRs) – another economiser.

The principle of mercury-arc rectification had been known since the turn of the 19th/20th century, although not then in common use. If excess energy was being returned to the line and not being absorbed by other cars, it could not cross the mercury-arc in the wrong direction and could cause the MaR to shut-down, rendering regeneration impossible.

Where but a few regenerative cars were operating amongst many more standard equipped cars in any section, the excess of high-voltage regenerative power forced back into the overhead was easily absorbed and used up with no difficulty arising in the disposal of the excess. However, where there were insufficient standard cars motoring within the section, to dispose of the regenerated power the excess would course back to the substation, or generating station, bus-bars feeding the line where some, or all, of a range of damage would occur. The most common damage comprised the 'blowing' of the electric light bulbs in both the regenerating car and also any other cars working in the same overhead section. Also, it had occasionally been found that tramcars lying at termini had their controllers go on fire as a result of excessive arcing of their controllers' fingers, especially in the early days of Phase One regenerative practice.

With mercury-arc sub-stations the rectifier is non-reversible and incapable of operation with inversion of the current as is the rotary converter. Mercury-arc rectifiers were non-existent in the First Phase of regenerative braking in tramcars. They were a fairly recent innovation by the time of the Second Phase. Indeed Glasgow Corporation Transport was recommending acceptance of the offer of Crompton Parkinson Limited for rotary converters and switchgear as recently as 1930.[6]

Here, the fact that MARs were irreversible and incapable of inverted operation meant that a different mode of dealing with the problem of overloading was mandatory. Resistor loading was the obvious answer in dealing with this condition. Schemes were developed to ensure that when a car in section was regenerating and there was no other car motoring to absorb the load, a resistor would be switched into circuit to take the surplus load and to ensure continuity of braking on the car concerned. The duration of such condition on any tramway system would have been extremely small, the lengths of sections being at half a mile, and the energy loss negligible. Such schemes of resistance-loading had been in use on railways where very much higher regenerative currents required handling but they had been found to be perfectly efficient for railway usage in dealing with their constant experience. To recapitulate, in summary, this was the recognised approach to such problems:-

Existing conditions of Supply	Change necessary for regen working
1. Shunt-wound rotaries without reverse-current relays	1. No modifications required
2. Shunt-wound rotaries with reverse current relays	2. Remove relays, or set them up to 40% reverse current
3. Compound-wound rotaries	3. Cut-out compounding. Relays, if any, as in (2) above
4. Mercury-arc rectifiers	4. Provide resistance load device to absorb excess current generated

All of this indicated that the problem of dealing with regenerated current did not present any more serious difficulties![7]

POWER LOSS IN FIRST GENERATION REGENERATIVE TRAMCARS:

In the First Phase of regenerative braking tramcars during the first decade of the 20th Century, many episodes of power loss in service were recorded against regenerative cars. With hindsight these were most likely to have been due to the additional field connections to the regenerative equipments having been poorly soldered to the existing wiring within the motors of modified prime movers and so subject, as mentioned before, to excessive road shocks from foreign bodies like stones wedged in the rail grooves and the hammer-blow of crossings, not to mention the abrupt jarring of car trucks negotiating tight-radius curves at fair speed. Inadequate soldering of wiring junctions would begin to work loose by vibration, spark and quickly burn through, jerking them finally apart at the next abrupt turn. Failures like this in the regenerative circuits were very dangerous as they rendered instantly total failure of the regenerative brake mechanism and the motorman would have no means of being aware of this. This led to many collisions, crashes and run-away cars.[8]

With the degree of advancement within the Second Phase regeneration practically all the equipments included purpose-constructed regenerative motors that eliminated the possibility of troubles such as these. The result was that accidents of this nature, all over, were eliminated.

PASSENGER REACTION:

From the passenger's point of view much pleasure was experienced from the increased speed and smoothness of acceleration and braking, but a certain jerkiness in motion, under inexperienced motormen, detracted from the ideal ride. It will be recalled that this had been especially commented upon in Edinburgh. The continual "popping" of the car's internal illumination was much deplored and Glasgow's solution, as previously explained, was to employ boys armed with boxes of spare bulbs to board cars and fit replacements as required. This could have been addressed on the regen cars themselves by the insertion of overvoltage relays. In operation these inserted a dropping resistance in the lighting circuit. Other (non-regen) cars would still remain vulnerable, however.

Glasgow's solution . . . was to employ boys armed with boxes of spare bulbs to board cars and fit replacements as required.

AN INTERESTING GLASGOW EXPERIENCE:

There follows a reproduction of an internal communication from the Deputy General Manager to the Chief Assistant Electrical Engineer, Glasgow Corporation Transport.

Typical terrain covered by Glasgow's Spiersbridge substation.
(Photo: J Venn)

```
Mr J.G.Boyne                                                    G.C.T.
Ch.Asst.Elect.Engr
G.C.T.
5/5    DMS                                                18th April, 1957

Regenerative Cars operating on Area
Fed by Mercury Arc Sub-station

      In 1938, when Spiersbridge Sub-station was commissioned, a service of regenerative
cars was operated in the area.  The sub-station fed from Giffnock to Cross Stobs, where
there are a number of fairly steep gradients.
      The area being an outlying one had for the greater part of the day a light load service
in operation, and it was possible for one Regen. Car to be operating without other cars
being present to take up the power generated.  Due to this condition, high voltages could
be present in the line and on the sub-station busbars.
      The high voltage present under those conditions damaged coils associated with the
DC circuit-breakers on the DC switchboard of the sub-station, and cars lying at termini
were also damaged, fire in one or two cases taking place.
      An attempt to overcome this condition was made by installing in the sub-station an
absorbing resistance controlled by voltage-operated relays which, when the volts
exceeded a pre-determined figure, switched the resistance across the station busbars.
Conversely when the voltage fell the resistance cut out.

                                    -2-

      Owing to space limitation, resistance was made "short-rated" and protected against
prolonged load by a thermal relay.  In practice the load to be absorbed when a Regen.
Car was feeding back into the line was of such magnitude that the thermal relay
operated, cutting the resistance out.  The interval of time required to reset the thermal
relay was such that the Regen. Car could again operate before the thermal protection
had reset to normal.  This condition made the use of the absorbing resistance useless.
      To house a resistance of continuous rating would have required an additional
building, which was not considered desirable.  Regen. Cars were therefore withdrawn
from the area.

      Copy to      C.A.E.E.                        Letter signed
                   File                            by
                   Extra                           Mr.W.L.McLennan
                                                   Deputy General Manager
```

THE SAVINGS UNDER REGENERATIVE BRAKING – AN EXCURSION INTO ECONOMICS!

Early in Phase Two propaganda was largely aimed at sales promotion of the new regenerative materials. This was before the bespoke, or custom-built regenerative braking motor came on the market and all tramcars had to be fitted with rewound to compound motors and their controllers modified for compound field control. It had been estimated that under the then prevailing conditions, the total savings that could be realised by conversion to regenerative braking were sufficient to pay back the initial outlay within two years. This had been estimated through the saving of current, (based on experience with the Manchester and Glasgow cars so fitted). Also taken into account were the savings in brake maintenance, electric maintenance and from the schedule speed having been increased, so resulting in a saving of 'platform charges. Furthermore, there was the saving as a result of the diminished fleet of cars required to cover the day's mileage on each route thus equipped. This reduced the overall maintenance of the cars and eventually of the fleet in general.[9]

Henry Watson, in *Operating Costs of Trams & Buses,* worked out the typical figures of savings that might be effected, thus:-

Saving in current:	£95/year/car
Saving in Mechanical Brake Maintenance	£6/car/year
Saving in Electrical Maintenance	£17/car/year
Saving in Platform Charges & Car Maintenance	
(bodies & trucks)	£91/car/year
Total Survey of Savings	£209/car/year

These figures would vary, naturally for different systems, either positively or negatively, but were representative of twelve large tramway systems for the period 1928-29, before regeneration.[10]

The Savings in Current:

In 1932, sufficient evidence had already been drawn from working regenerative tramcar equipments to indicate 25-30% was saved by regeneration, as compared with standard equipments under equivalent conditions, with the exception of hilly routes and routes with overall high-schedule speeds, where the saving became very considerably greater.

The cost of current per mile averaged over the twelve mentioned undertakings had been taken as 2.07d and, taking 100 miles a day as a reasonable average mileage over the 365 day year, the cost of power was then £315 per car. The 25-30% power saving was measured at the car, itself.

Trolley-wire losses on account of the smaller amount of current supplied from the sub-station, or generating station, showed a certain saving as did the regenerating cars themselves, being equivalent in their regenerative activity to powerhouses or sub-stations, thus generally diminishing the distance the current would be required to go to the nearest car taking power. Thus, trolley-wire and track losses would be reduced from approximately 10% to 5%. This saving in power would then approximate to £95-£110 per car per annum. To accept a conservative estimate, the lower figure had been used in the Metro-Vick Descriptive Leaflet.

'... sufficient evidence had already been drawn from working regenerative tramcar equipments to indicate 25-30% was saved by regeneration, as compared with standard equipments.'

The Saving in Mechanical Braking:

Braking dissipates quite the most power of all the tramway functions so that the reduction in power consumption through regenerative braking was directly reflected in the wear and tear of braking equipments. Constant braking from high speeds required the frequent renewal of brake shoes, of any type. The attrition at high speeds was very much greater than stopping from a speed of, say, only a few miles per hour. Regeneration takes the entire control of deceleration to about 5 or 6 mph. without the use of any mechanical, or other, brakes, thus relieving the brake-wear enormously. It also relieves the constant acceleration and braking resorted to when trailing behind other traffic with the ordinary type of equipment. The £6 saving per car, per year, claimed was indeed a modest one!

The Saving in Electrical Maintenance:

Controller maintenance was much reduced, being subject to much less wear and tear because for a great part of the time, they are only controlling the relatively small shunt current to the motors and were not required to break the large motoring currents of the standard cars controllers when tagging along in close traffic conditions.

'There was no difficulty in the increase of schedule speed by ten percent with regenerative equipments'. Glasgow Corporation Transport would dearly have loved to have done just that by eliminating their remaining slow cars like 968 forced to operate in the area covered by their Spiersbridge Mercury arc rectifier.

(Photo: FNT Lloyd-Jones/ OnLine Transport Archive)

Despite this saving, however, the conversion of existing equipments necessarily involved a very complete overhaul and rebuilding of the car's whole electrical equipment. Part of the cost of conversion was this overhaul which resulted in practically new equipment when completed, so that over the period in question, electrical maintenance would be very small indeed consisting mostly of standard inspection only. The average electrical equipment maintenance cost was 0.22d per car mile and a very modest claim for the regenerative vehicle would have been that half of it was saved in the first two years resulting in the £17 figure, per year.

THE SAVINGS FROM & INCREASE IN SCHEDULE SPEED RESULTING FROM REGENERATION:

Importantly, regenerative equipments enable an increase in schedule speed to be made for frequent-stop routes without having to pay for this increased current consumption. To obtain an increase in schedule speed the brakes require to be applied at that higher speed, and as energy thus dissipated in the brakes increases as the square of the speed, every small increase in schedule speed increases the power consumption to an alarming degree. This does not occur with the regenerative equipment as the braking energy is returned to the overhead at quite a reasonable efficiency. Current net consumption is comparatively little affected, in a regenerative scheme, by altering the schedule.

As such, regenerative braking was a new tool placed in the hands of the traffic manager whereby he could meet public demands for improved schedule speed and at the same time save money on it! There was no difficulty in the increase of schedule speed by ten percent with regenerative equipments that would reduce the platform charges per mile by a similar percentage. What happened was that a given daily mileage was run by fewer cars. Not only would the wages of motormen and conductors be reduced per car mile, but also the cost of inspection, car washing, and so forth!

As an example: take a 7-mile route with a 10-minute headway between cars. The schedule speed of the cars would be 9 mph. Ten cars in service would give 3¼ minutes waiting time at each terminus. If the schedule speed was increased to 10 mph, the service could be operated with only nine cars.

The platform charges on a tramcar were about 4½d (1.9p) per car mile and car body and truck maintenance about 1½d (0.6p) per mile. On these a 10% saving would be equivalent to a saving of £91 per car per year. This speeding-up of cars was an extremely attractive proposition but had hitherto been impracticable owing to the prohibitive amount of power consumed in braking when ordinary equipment was used. With regenerative equipment this energy was not wasted but returned into the system. On a certain hilly system where the car-miles per day were over 110, the saving indicated from all sources might total as much as £300 per car per year. This was no mean consideration! [11]

What price that tramway system was Halifax?

Halifax car No.11 on route 6 to Sowerby Bridge. The Corporation bus in the background has the distinctive 'Camelback' roof, lowering the overall height without infringing Leyland Motors' patents.

(Photo: HN McAulay collection)

REFERENCES – CHAPTER 7

1. The Tramways & Light Railways Association: Special "Brakes" Number:1908: No.69: Regenerative Control: pp 818-820
2. Webb, J.S.: Black Country Tramways: Vol.1 (1872-192): 1974, published by the Author
3. Groves, F.P.: personal communication: 15-6-01
4. Metropolitan-Vickers Company Ltd.: A New Development in Tramcar Control Equipment: Descriptive Leaflet No. 580/1-1: Form 1511.
5. Pilcher, R.S.: Miscellaneous News & Notes: The Tramway & Railway World: 1930 – November 13th: LXVIII: p 311
6. Tramway & Railway World: 1930-November 13th: p 287
7. Metropolitan-Vickers Company Ltd.: Regeneration and Substations: Descriptive Leaflet No. 580/4-1: Form 1511: Reprinted from The Electric Railway, Bus & Tram Journal: 1932 – November 11th

8. Markham, J.D.: personal communication
9. Metropolitan-Vickers Company Ltd.: Descriptive leaflet No. 580/2-1: Form 1512: Reprinted from The Electric Railway, Bus & Tram Journal: 1932 – October 14th
10. Watson, H.: Operating Costs of Trams and Buses: The Electric Railway, Bus & Tram Journal; 1931 – April17th: quoted by M-V Descriptive Leaflet No. 580/2-1.
11. Metropolitan-Vickers Company Ltd.: Descriptive leaflet No. 580/1-2: Form 1512: Conversion of Standard Tramcar Equipments to Regenerative Braking: Reprinted from The Electric Railway, Bus & Tram Journal:1932 – October 14th

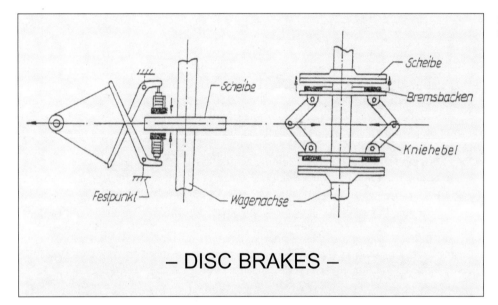

– DISC BRAKES –

Forms of early disc brakes. (Drawing by Rev. David Tudor).

Woltersdorf, near Berlin, September 2000 with wartime utility tram (KSW) train. The braking solenoid connectors are visible as Motor Car No.7 turns round matching trailer No.22. *(Photo: Rev. David Tudor)*

THE CONTINENTAL PRACTICE OF TRAMCAR BRAKING

The Reverend David Tudor examines how tramcar braking was developed on the continent during the periods covered by this book, and takes a look at the future.

In the USA, at the turn of the 19th/20th centuries, the compressed air brake was almost universally adopted. Tramcar controllers were not equipped with brake notches and this can be seen today on the Third Avenue Car, 674, at Crich. Contemporary developments on the European mainland took a different direction. As early as 1899 the firm which became Siemens Schuckert was supplying controllers with the so-called short-circuit brake – the name deriving from the feature that the single electric brake notch quite literally put a short circuit across the output of the traction-motors switched into the generating mode. This is the equivalent of the more modern controllers' last electric brake notch. As speeds increased this feature was refined by using the starting resistors to provide a graduated electric brake – the resistors being gradually cut out through five or six notches until the final brake notch was reached and the short circuit was applied. This enabled a smoother stop to be achieved. The employment of the resistors, or rheostats, led to the term 'rheostatic braking' being applied.

In the early years of 19th century tram and trailer operation was common practice in continental European cities. In many places the best cars from the horse-tram fleet were adapted to work as trailers behind the new electric cars. (Some of these even survived to become museum pieces after their passenger carrying days were finished). Later, new types of trailer cars were introduced to replace the small ex-horse trailers and it became apparent that some form of trailer braking was necessary, especially if service speeds were to increase. Many interurban tramways using two or more trailers employed air-brakes to provide braking over the entire tram-train. However, urban tramway operators regarded the air-brake as an expensive alternative to the reliable rheostatic brake which, using the traction motors, incurred no additional costly equipment such as compressors, air tanks, cylinders, brake valves and pipework. Furthermore, the electric brake did not freeze up in winter as air-systems frequently did!

In 1923 Berlin tramway engineer Eduard Kindler designed a new type of trailer car of lightweight form without a separate truck. To improve the braking effectiveness of the total of 803 cars built of this class, he introduced a form of disc-brake mounted on the axles. It was soon discovered that disc-brakes – needing relatively little movement to apply them compared with older tramcar brake-rigging using cast-iron brake shoes – were ideal for use with a solenoid fed from the braking-current of the motor car. Another advantage was that application of the electric rheostatic-brake caused the solenoid brakes on the trailer(s) to operate proportionately with the motor cars' braking performance. A harsh application would cause a harder application of the disc-brakes, thus giving an equal retardation along the entire tram-

. . . urban tramway operators regarded the air-brake as an expensive alternative to the reliable rheostatic brake

Above: a Freiburg 3-car articulated 4-bogie single-ended tramcar, seen from the non-boarding side on 28th May 1993. *(Photo: Struan Robertson)*

Some 13 years later these cars are still in all-day front line service, but the centre sections have been rebuilt to low floor configuration giving easy access for wheelchairs and push chairs, as well as making for easier boarding for the less mobile. This view was taken in October 2006. *(Photo John A Senior)*

train. Because of the time taken for the magnetic effect to build up and subside, the trailer brakes would bring the set to rest 0.5 seconds after the braking current ceased. In this way, the traditional continental three-car tram-train would usually come to a complete stop, it being necessary only to apply the hand-brake on the motor-car to hold the train stationary. By 1930 this type of brake became a standard feature across Europe. Even 1960s' Duewag modern trams had this feature, not only on trailers but also on the non-motorised axles of the trucks under the articulations of long two- or three-section articulated tramcars.

Close inspection of photos of traditional trailer-car couplings will reveal – in addition to the mechanical coupling and the thin cable for lighting, etc, – a very thick cable. This is part of the electrical connection for the brake-solenoids. The return path is via the earth return (ie the track). As the large plug and socket for the brake connection could be parted in the event of a coupling failure, it was important that this did not lead to a catastrophic failure of the most important form of braking – the rheostatic brake. So the Brake Protection Resistor was fitted on all motor-cars and, being wired in parallel with the trailer brake solenoids, was able to maintain an alternative current path should the couplings become defective.

For the sake of completeness it should be mentioned that the rheostatic braking current, as well as being used as described above, is also dissipated in the resistor grids and can be switched through the car heating units in winter. It can be used not only to activate track brakes (as in Glasgow and LCC) or auto-sanders (to drop sand under the wheels), but also even actuate a magnetic valve (in a form of interlock brake protection) to dump the air from the air brake cylinders when the motorman applies the rheostatic brake. This can be seen on MET Car 331, also at Crich).

The other form of electric braking – regenerative braking – as will now be appreciated, has the facility of sending the current produced by the traction motors working in braking mode back into the overhead wires, thus reducing the amount of energy required from the normal electrical supply. The prospect of reducing operating costs is high on any tramway manager's agenda on the continent, just as in Britain. Faced with the immense loss of patronage as a result of the World economic slump in the years 1929-1936, the management of the Aachen tramways, having lost 45% of their passengers, were keen to cut their costs. The Director, Marcel Cremer-Chapé, and Chief Engineer, Peter Schings, developed the so-

A modern Combino modular articulated tram, No.765 in Berne. These have not been without their own problems since entering service but were structural rather than electrical.

(Photo: Tom Robinson)

called 'Aachener Nutzbremssytem' (Aachen's useful braking system) patented on 4th March, 1932. This fed the braking current back into the overhead wire for use by other trams. A demonstration was staged on the tramway between Burtscheid and Siegel. After positioning two trams at the top of a hill and one at the bottom of the same gradient, the current was switched off at the sub-station. The two trams using their regenerative brakes moving downhill provided sufficient power to propel the third tram up the gradient.[1] It was claimed that over 20% was saved in the traction current used on the system by fitting regenerative brakes to 93 cars – almost the entire Aachen fleet.

Only in the last twenty years with the advent and development of Alternating Current traction systems and power electronics, has regenerative braking become almost a standard feature of modern electric traction as used on Light Rail. On the South Yorkshire Supertram system the motorman's Voltmeter can be seen registering almost 1,000 Volts when a tram's braking current peaks.

That there is nothing new under the sun is proven by articles published recently in a German tramway periodical. It reported a research project initiated by the French tramcar builder 'Alstom' in Alsace, in which a former Karlsruhe articulated tram has been adopted to provide regenerative braking-current to power a flywheel. This is the size of a cupboard and is mounted on the tram. The flywheel provides traction current for acceleration whilst its speed of rotation is boosted at major stops by a short stretch of overhead catenary. The objective was both to recycle braking energy and save costs and visual intrusion by only having short sections of trolley wire. A second report in the same magazine concerns a modern variation of the old 'stud-contact' idea! Again in France, lengths of track have been installed in Marseilles to test the concept and 8.5 km of new tramway has been constructed in Bordeaux, partly using this method of current collection. Instead of studs, contact rails are let into the road surface in the centre line between the rails (like LCC conduit) and are cut into 8 metre long sections, separated by 3 metres of dead contact rail. Only two 8 metre lengths can be energised at once and the long articulated trams switch the rails on as they pass over sections. As a precaution, traction batteries are being fitted to the trams! Like the conduit system of old the cost of this system is many times that of the conventional overhead wire system. One wonders, too if the same precaution will be implemented as was found necessary on the stud-contact system of old, whereby a wire-brush bonded to earth was fitted to the rear fender of each tram. This was designed to fuse any errant studs that remained live before any unsuspecting horse trod on a live stud with fatal consequences!

REFERENCES – CHAPTER 8

1 Source "Strassen-& Stadtbahnenin Deutschland Vol.7" Holtge and Reuther. 2001

... Only in the last twenty years with the advent and development of Alternating Current traction systems and power electronics, has regenerative braking become almost a standard feature of modern electric traction

A modern LRT built by Alstom operating on the APS system of surface-contact collection in the old city area of Bordeaux shortly after this part of the system opened in May 2004. The absence of traction poles and overhead wires is very noticeable.

(Photo: John A Senior)

Mulhouse, a town of Museums including the French National Railway Museum, is one of the latest European locations to have installed a modern tramway system, though in this case more conventional than the Bordeaux one opposite. Twenty-six Alstom Citadis cars are used and there are plans to link Mulhouse with Strasbourg using the tram-train principle successfully working in Karlesruhe where the LRVs run over the main line railway network. The rapid acceleration, and deceleration through the regenerative braking, makes such ideas feasible. This is the terminus of Line 2, seen in October 2006, some 4 months after the system opened. High rise flats in the background give a hint as to traffic potential. *(Photo: John A Senior)*

THE THIRD PHASE

OF

REGENERATIVE

BRAKING —

Whereas previous attempts at regeneration were probably ideas ahead of their time, let down by the available technology, the widespread adoption of electronic control in the 1960s changed all that. Continental development is touched upon reflecting its influence on current British practice.

INTRODUCTION (BY STRUAN ROBERTSON)

It is not our intention to project this book formally into the Third Phase of Regenerative Braking. This is so wide a sphere of activity as to demand treatment of a more complete nature in the fullness of its ongoing accomplishment – it is a different world in transport! A few words of introduction to what is a most interesting sphere of engineering concept, however, might not be out of place.

As previously mentioned, the early 1960s saw the demise of the first generation British Tramway Systems apart from Blackpool and the Isle of Man. Thereafter the ascendancy of the motor bus held sway, and still does. Germany, however, used the 1960s to modernise many of her already excellent systems while France, in 1975, began to follow suit in the contemplation of what should arise as the New French Standard Tramcar. Lille, in 1980, and Toulouse a few years later, launched into modern

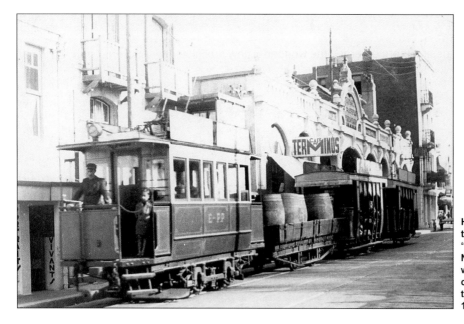

Here is a typical pre-First World War type of continental tramcar – the "Tramway d'Etaples a Paris Plage. Car No.1 is sporting three trailers – a mineral wagon and two open passenger coaches; a heavy load for so small a tram! This was photographed in July 1939 by Struan Robertson.

Boulogne No. 104 is an early post World War I tramcar featuring perhaps the ugliest vestibules that were quite out of keeping with the neat lines of the saloon. This is Wimereux also in July 1939.

(Photo: Struan Robertson)

articulated tram-trains, followed by increasing interest elsewhere in the larger cities of the European Continent: Nantes in 1985 and Grenoble in 1987 were followed by increasing interest in the Western countries, and momentum is still gathering.

The system of single-decked tramway trains had become endemic on the continent. From the beginning, in the late 1880s, small single-truck cars were the order of the day with rebuilt horse-cars as dummy trailers tacked on at rush-hour periods.

Cities grew as industry drew the populace of the countryside with the promise of work in mills and factories, and greatly increased city transport became imperative. The day of the long, double-bogie car dawned. Throughout the inter-war period of 1919-1939 these held sway. A small number of continental cities did use double-decked tramcars but not really to any great extent. Elsewhere in situations where tramway systems survived the onslaught of the motor bus, the major concentration seemed to be in terms of greater comfort for the tram-travelling public and the enhancement of remaining services, but the tramcars, and their trailer passenger saloons were getting larger and larger.

It took the electrical revolution of the late 20th century to produce solid-state electronics before electric transport took a quantum leap forwards in such gigantic proportions. From every conceivable aspect related to control, motion and general design there emerged the wonderful new articulated tramway-trains of the turn of the millennium. With this was the coming of age of Regenerative Braking amongst the many other attributes.

Following World War II, with its lack of almost all modern materials, depressed economy and state control of commodities, the stage of the long-bodied bogie car-train remained static until industry gained momentum. These cars invariably had high loading platforms, usually needing two substantial steps up, relied on series-wound dc motors and coupled trailer carriages. With the coming of sold-state electronics and greater control, design went through a sea-change in the wake of social consciousness to low-loading platforms and better, anti-jerk safety grips and handles. Articulated coaches replaced those with couplings. This in itself eliminated the waste of time resorting to depots for the attachment of trailer coaches. Gradually, following experimenting with, and adoption of more modern motors and improved passenger comfort from every aspect, the Millennium had arrived!

The Solid State Electronic revolution allowed regenerative braking at last to come into its own with the elimination of all previous drawbacks. How far ahead of his time was John Smith Raworth? At least a minimum of seventy years! Low

voltage alternating current came into use and commutator-less motors found their way into transport. The long-bodied car, latterly space-occupying on the street curves, developed into multi-modular units, ultimately with the elimination of undercarriages by hanging them from short motorised modules interspersed along the car length. These vehicles were ideal for snaking their way through narrow and crooked streets and could handle curves with a minimum radius of 18m. Recent developments in maximising low-floor accessibility have seen the introduction of the vertical motor-cum-suspension drive unit housed over a single driving wheel between each of two units of a typical five-module articulated tram-train. More and more modern principles are being tried out all the time, such as self-aligning wheels that markedly reduce the screech of wheel-on-rail and the grinding displayed, even on modern cars, on tight curves. The next generation of British trams, if it is ever allowed to blossom, will benefit from all these innovations.

Top: In the period 1919-39 the tramcars were getting larger and larger. This is Vicinal No. 9825 in Bruges Square immediately before the outbreak of war in 1939.
(Photo: Struan Robertson)

Lower: Post World War II Zurich bogie cars had modern streamline bodywork with roof-mounted resistances but they were still of the pre-electronic period. Here is No.1386 with trailer cars 634 and 640 on Route 13.
(Photo: Struan Robertson)

Development of the Continental Tramcar

As a lead-up towards twenty-first century continental tramway practice and the development of the third, and even fourth stage regenerative control there is merit in taking a quick glance at a few representatives of what went before. This will also allow comparisons to be made with parallel practice in the United Kingdom prior to the virtual demise of its first generation tramways.

A similar situation existed on the continent whereby the big towns and cities progressed more rapidly. Smaller systems, without the background of a strong bank balance, tended to make-do and mend.

The two World Wars brought in their trail fairly rapid advancement in traction technique and the inevitable expense hit the smaller towns and rural systems which largely went to the wall. Many of the large intermediate systems, however, held on tight to their tramway systems and here was to be found the rapid advancement of the electronic age. The rapid rise of the internal combustion engines did lead very many of even the greatest cities to abandon their tramways. Paris is an excellent example and this was despite their advanced theories and successes with regenerative braking as described earlier. Their system was scrapped, entirely, in 1938.

However, elsewhere, the pressure of demand to move heavy populaces in towns with narrow, and sometimes, tortuous streets centrally where rapid peripheral development was taking place, led the thinkers to realise and exploit the suitability of maintaining the tramcar in its now modern style as the main people-mover.

TRACING THE ELECTRICAL ENGINEERING DEVELOPMENTS
(BY JOHN MARKHAM)

It was in the early part of the 1960s that electronics began to swing away from using thermionic valves to semiconductors in communication and instrumentation. This was largely as a result of developments in the understanding of the physics of the processes taking place at semiconductor 'junctions' within the doped crystals they comprised, and the parallel developments in the technology required to 'grow' them. The specialised 'chips' that now form part of everyday items have evolved from those early days.

The 'power' of a computer chip derives from micro-engineering on a large scale, both in terms of its design and the precision technology required to make it. A single 'chip' may now contain over ten million transistors and the largest items are the points of connection to the outside world.

However, whilst one can marvel at what is achieved in terms of calculation, data handling, command and control, entertainment, etc, such 'chip' devices cannot do physical work. Substantial powers are required for this and we still need the translation of power into mechanical form by electric motors of one type or another.

Macro-engineering of semi-conductors has been required to provide static means of electrical power control. The challenges are different. Instead of the device having a vast array of differences for the multifunction logic applications, the necessity is the provision of great uniformity both of the substrate and of those few 'junctions' that do exist.

It is not the purpose of this text to go any deeper into the science and design of power semiconductor devices, but it is hoped the above rather simplistic description will be helpful to the reader to understand why only a limited range of power semiconductor devices exist. Their application to power modulation for motor control requires careful and specialist techniques not found in what were more traditional ways of achieving similar overall results. Electric traction nowadays is an important application of power electronic techniques. The most basic of all power devices is the diode. In its most simple form, such a device conducts electricity in one direction but not in the other. Arrays of these are used to form, for example, substation rectifiers, providing dc for traction from a three (or more) phase ac supply.

The next device of significance for traction applications is the thyristor. This acts as a fast switch, which can be triggered from OFF to ON. That is all it will do. For it to revert to its OFF state it has to be starved of current for a minimum period (the turn-off time) by other means. More versatile is the gate turn-off thyristor (GTO), which, as its name implies, can be switched from ON to OFF as well as OFF to ON.

There are other devices as well, but these are the main ones used in power control of electric traction vehicles.

Any power control system uses a configuration of such devices in conjunction with reactors and capacitors to form a chopper, inverter, or other such composite power modulator that is required.

First used on battery vehicles such as milk floats, choppers have, for about three decades, been widely used on fork-lift trucks and similar applications. Tramcars, trolleybuses and trains have followed. All used ac commutator motors, series wound initially, but later some applications used separately excited fields whose control, by a separate power modulator was integrated with that of the armature. Separate, but integrated, control of the motor armature and field – was that not what both Raworth and Johnson-Lundell were doing at the beginning of the 20th century?

For any form of chopper to be used we have to exploit properties of electric circuits that have so far been ignored in our dealing with switched resistance control. To an electrical engineer this is the study of transients; the effect on a circuit of operating a switch (ON or OFF).

Closing a switch is chiefly used to connect a source of electrical energy to a load so that work may be done. However, at the instant of closure the current does not rise to its final value immediately. The voltage of the supply is divided amongst the various elements of the current in the ratio of their inductances. The initial rate of rise of current is then governed by the sum of all the inductances in the circuit. As the current starts to flow, however, voltages also become established across resistive elements in accordance with Ohm's Law, which oppose the voltage that has been applied. Resulting from this the rate of rise of current decreases. A simple analogy of the effect is that of a ball float valve in a cistern or tank that gradually shuts off the supply of water as its level rises.

A fundamental feature of a chopper is to switch off again almost as soon as it has been switched on, and before the current flowing has had sufficient time to build up beyond that required to be passed into the load. By repeating this process at a high rate – usually several hundreds of times per second – and increasing the ratio of ON time to OFF time, a power modulator is created that can provide a variable voltage, with the current in the load limited to whatever is required.

It is not appropriate in a general work such as this to go into details of the many chopper circuits which have been developed over the years any more than popular books on steam railways go into the design of valve-gear. Nevertheless, here there is another analogy. Armature control by chopper on starting is similar to opening the regulator in full gear, and field control by chopper whilst running is almost the same as 'linking-up' with the cut-off of the valve gear.

It will be appreciated that field strengthening at speed will cause the armature current to fall and the armature voltage to rise. These are just the circumstances that prelude regenerative braking. There is a snag, however. The early choppers were uni-directional. An alternative path for the current had to be provided. A diode, or reverse conducting main thyristor, would help here, though regeneration on this basis would only work down to 'base speed' – the full-voltage characteristic of the motor. Below this speed the armature voltage would be less than that of the line, and power could not be returned.

Or could it?

In motoring, the chopper acted as a variable ratio dc transformer as far as the motor was concerned. By reconfiguration of the chopper-motor circuit it was possible to 'pump-up' the voltage from the motor to match, or exceed, that of the line so that return power flow could take place. The analogy is the water injector for a steam locomotive boiler.

But what if the electrical supply was of limited receptivity, or could not accept the returned power at all? No problem. Line voltage detection on board the vehicle sensed this and yet further action by a chopper diverted the surplus generated power to an on-board loading resistance in which it was absorbed and dissipated. In modern vehicles these are often mounted on the roofs.

Most, if not all, tramway applications tended to favour the use of series motors rather than the separately excited scheme outlined above, and were able to avoid the need for a field chopper. Some clever circuits were developed to enable one chopper to suffice.

Further developments in control techniques brought in the use of automatic wheelspin and wheelslide detection and correction that was integrated with the traction package.

For many years the Holy Grail was the elimination of the motor commutator.

For many years the Holy Grail was the elimination of the motor commutator. A series or separately excited traction motor had a commutator and brushgear requiring regular inspection, even if not actual maintenance. Then there was the ever present possibility of a flash-over at brushgear that could do varying degrees of damage.

The induction motor required no commutator or brushgear. This allowed the active part of the motor core to occupy a greater proportion of the available frame size and needed only three leads for power supply. This was perceived as attractive even without going to a linear design where the rotating machine could be eliminated altogether.

The accommodation of induction motors necessitated a further development in power control technology. Choppers on their own could not achieve what was required. The three phase induction motor demands a three phase supply which has to be synthesised. For a variable speed application, such as traction, the supply has to become three phase with simultaneous variable voltage and variable frequency. The apparatus to produce this is entitled an 'inverter', and can be considered to be, in crude form, six choppers, two or three of which are on at the same time. So, as well as the variable on-off ration we had with the simple chopper, we now have choppers operating in a pre-determined synchronised sequence as well, and at a varying rate. Reversal of the sequence performs the function of reversal of rotation of the motor (saving a reverser). Further complexity arises for dynamic braking (regenerative or rheostatic) in that induction motors do not self-excite, and have to be provided with their magnetising current from the inverter while simultaneously injecting power back through the system. This is not a topic for detailed analysis here.

An important feature of electronic power control has been the integration of dynamic (regenerative and rheostatic) and friction brake control. This is often done in conjunction with the wheelspin and wheelslide detection and correction equipment. By this means the level of deceleration demanded by the motorman is achieved in a priority hierarchy which gives preference to regenerative, then rheostatic, supplemented by friction braking first on un-motored axles followed by those on motored axles. The proportions are not under the control of the motorman but are determined by the control gear to suit the vehicle speed and rail conditions. In extreme conditions the above may be assisted by the use of magnetic track brakes and the application of sand to the rails.

Sheffield Supertram 105 in its first Stagecoach livery. These cars are presently undergoing a mid-life refurbishment.
(Photo: IG Stewart collection)

Many of the above features are built into the new generations of tramcars serving Manchester, Sheffield, Wolverhampton, Croydon and Nottingham.However, the induction motor's characteristic is such that its maximum torque cannot be delivered easily at rest. It has to be driven very hard, electrically, because at rest it has infinite 'slip' whatever the supply frequency. In many industrial applications such motors are often started off-load and are only connected to whatever they are driving after they have been run up to speed. The speed at which they run on load is then slower than that which would synchronise with the frequency of the mains feeding them. This difference is known as the 'slip' and is fundamental to the ability of the induction motor to produce any torque. This 'slip' is typically in the range of 3-10%.

The induction motor is not the only form of commutator-less motor. Extremely satisfactory performance was achieved with a new tramcar fitted with four switched-reluctance motors in 1985-6 and operated at Blackpool. The initiative for this development lay with GEC Traction Ltd., of Trafford Park, who sponsored the tram (which ran as '651' in Blackpool). The basic technology of the static control of the variable speed of such a motor was that of Switched Reluctance Drives Ltd., a technology company of the University of Leeds. Such a motor works without 'slip'. The controlling electronics are synchronised with the angular position of the rotor, much in the way that the electronics of present day motor cars time the firing of the cylinders. It can also produce its maximum torque at standstill, just like the trusty series motor, and is capable of maintaining a locked rotor should this be required. (But would anyone want to use a traction motor as a holding brake?)

The reluctance motor – known as a stepper motor in light applications, has no squirrel cage and requires no winding at all on the motor, which is a salient pole construction In terms of controllability I know of no equal. On more than one occasion I demonstrated moving the tramcar, all 17 tons of it, from rest by 3/8in (10mm), in two increments, each of about 3/16in (5mm), and without operating it against the brakes.

Croydon Tramlink cars 2534 and 2549 at the Lebanon Road tramstop on 3rd May 2001. *(Photo: BJ Cross)*

PART TWO

TECHNICAL CHAPTERS

10

THE ELEMENTS OF RAILED STREET ELECTRIC TRANSPORT

Before venturing into the technology that brought about Regenerative Braking, it is necessary to go back to basics to appreciate first of all the working and theory of the conventional series-parallel controller and the various other means of braking of the traditional tramcar at the motorman's disposal.

Fundamentals:

First - The Electrical Engineering Short-hand!

In the complicated series of representative symbols used in transport electrical engineering we come across:-

Motors: —(A)—mm— Indicative of its Armature, together with its Field System.

Switches: or —o o— Switch in normally open circuit contact

or —o─o— Switch in normally closed circuit contact

(No.) Normally open contactor switch

(Nc) Normally closed contactor switch

Contactor switch with B.O. (Blow-out) coil - open

Contactor switch with B.O. (Blow-out) coil - closed

—| |— Previously represented any switches in older schematic drawings - (open)

—●— Previously represented any switches in older schematic drawings - (closed)

Resistors: ⌐⌐⌐ Resistance - non-inductive (Power Resistance)

—W— Resistance – (Control Circuit)
—m— Coil winding

Wiring: ┼ Crossing wires NOT connected (i.e. never connected) – Note: never put a dot on the crossing – it is never done!

┷ Crossing wires that ARE connected

⊥ Earth (or:- ⊥⁄⁄⁄)

These are adequate to be going on with; the rest may be picked up in transit.

The Basic, Fundamental Power Schematic:

In the first phase of electric tramcars by far the majority had their motors wired in what was known as the Short-Circuit, or Shunt Transition mode. This is shown, in the following drawings and was the simplest and most rugged method of all times. Remember these fundamental circuit diagrams for they hold good for almost all of the first phase cars.

(A) Neutral: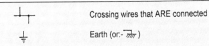

In this simple diagram the resistors and all secondary circuits and connections have been left out. Perhaps, also, a reminder that the jargon regards as "Closed" circuit as when it is an intact, closed ring, or "On" in homely parlance, i.e. complete, live and functioning, and "Open" when switched off, or interrupted in any way. Picture a day-breaking depot scene. When the power is switched on from the overhead (O/H) by morning staff closing both No. 1 end and No. 2 end canopy switches, the car is at rest because the traction switches M.S and G are open, or neutral. The lights will have come on and the compressor will have started working (on an air-braked car) when switched on because their circuits are tapped off the T-cable, or trolley cable, above the canopy switches. The car is now ready to move off.

(B) Series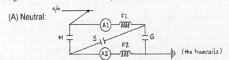

When the motorman moves the controller handle on to the first series notch and simultaneously releases the parking brakes, a train of events begins to occur. First, if there was a "line switch" (L), this would be first closed, followed by the closing of switch "S" in the initial stationary switch-on transient, allowing the current to build through the full resistor bank, and thence through A1, F1, the switch "S" and A2 and F2 to ground, on the tram rails, and thence back to the substation. The current flow through the traction circuit and full bank of resistances causes the line voltage very momentarily largely to dissipate before its effect on the motors can be appreciated. As the controller handle approaches the first series notch the

144

current is establishing itself. Most of the voltage drop takes place across the starting resistors as the motors have low resistance in comparison. As the armatures start to turn their rotation within the now magnetised main field poles of the motor carcass, or casing, begins to induce a voltage in the armature windings in the direction, or polarity, opposing the applied motoring voltage from the overhead. This voltage is called the "Back e.m.f." or electromotive force, and its production is a fundamental "Law of Physics" feature of electrical machines which will have a great significance in pages to come. The car would be gently lumbering towards the depot door as the current establishes itself in the first series mode, as in the second diagram. Trace out the travel of the current from the overhead through both motors to earth remembering that, although not drawn in for the sake of simplicity, the resistor banks are in full use, but will be cut out as the motorman notches up.

As the armature speed rises, so the Back e.m.f. increases, opposing the line voltage, which meant that the voltage available to force current through the circuit was reduced and in consequence, the current was reduced. Motor torque, and hence accelerating tractive effort of the tram, depends upon the magnitude of current through the machine. Although not an absolutely linear relationship when using series motors, this is acceptable for present purposes. With the fall in current as the car accelerates, the rate of acceleration fades and, in order to reinstate it, the motoring current must be increased to its former value. By cutting out a section of the starting resistance the voltage drop across that section is eliminated which allows the line voltage to be redistributed, giving a greater share to the motors. This process is repeated through the series notches on the controller until full series is achieved. On this notch the two motors divide the entire line voltage between them approximately equally. No two motors are exactly alike and their speeds may be slightly different due to differences in wheel diameter and so forth. The Back e.m.f.s will be less than the individual motor terminal voltages as some line voltage will still be accounted for by the voltage drop due to the current through the inherent resistance of their armature and field windings. At this stage all resistor segments are out of circuit, the car is running at half speed and no current is being wasted. This is the first of two, only, economical running speeds of the standard series-parallel tramcar. It may travel any length of time on this notch at a reasonable half-speed for city centre congestion without wasting any power, and is known as the first "Running Notch". The full series notch is not fixed speed. It is variable, and conforms to the half voltage characteristic of the motors. This is a characteristic allowing the speed to rise with the tractive effort decreasing. The car can therefore accelerate on that notch, which it does, until the tractive effort has fallen to match the resistance to motion (from gradient, friction, windage, etc.) for the location. Only then does speed become constant.[1]

The Transition Phase:

We now come to the Transition Phase, between top series and first parallel notches, in what is called "Short Circuit", or "Shunt Transition". There are four main stages in transition:-
1. Re-insertion of series resistance
2. Short-circuiting No. 2 motor (when it de-magnetises)
3. Opening circuit of No. 2 motor
4. Connecting No. 2 motor in parallel with No.1 motor
   And there can be a fifth –
5. Reducing the series resistance.

(C) Transition

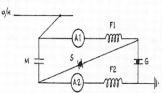

Deliberately neglecting the transition resistors in these diagrams as not necessary, this diagram shows switch "S" being closed, as also switch "G", the latter short-circuiting No. 2 motor, which now de-magnetises.

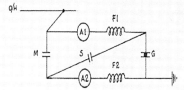

Switch "S" is opened, open-circuiting No. 2 motor momentarily. No.1 motor now takes the full line voltage while No. 2 motor is idle, in the second stage of transition with the car being driven by No. 1 motor only.

(D) Parallel

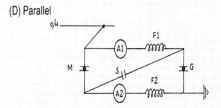

Switch "M" is now closed, re-energising No.2 motor and throwing the motors into parallel. The car is now running on both motors in the first parallel position and the series resistances will have been brought into play to use up a portion of the line voltage during transmission, and were reduced after full parallel was attained.

Taking the above black dots of the skeletal motor circuits in the order of these diagrams and setting them in exactly the same order, in tabular presentation, one gets what is known as the "Dot Chart" – a form of short-hand representation of the sequence of switch routine. Sometimes the dots are replaced by crosses.

The following four photographs taken by Struan Robertson illustrate the stages in rewinding B T-H armatures.

Stage 1 – completely stripped down

The "Dot Chart" Representation of the Sequence of Switches:

In the "Dot Chart", of great fondness of visual demonstration of the sequence of switch action, and met with in almost all power schematic diagrams, errors of representation may sometimes arise. For absolute accuracy of definition without possibility of failure or misunderstanding, the "Bar System", indicative of the exact open-or-closed situation of all switches at any one moment, is more fully to be relied upon. The "Dot Chart" representation of the above sequence of switches would be

	M	S	G
Series		●	
1st Transition		●	●
2nd Transition			●
Parallel		●		●

From the Dot-Chart one quickly perceives the outline of the Bar representation to follow and at this stage there seems little difference between them. In practice, however, dot representation can be capable of misrepresentation which the bar principle avoids.

The "Bar Chart" Representation of the Sequence of Switches

Here, all of the detail of the group of individual drawings, above, is condensed into the following single, simplistic, figure. It acutely accurately indicates the detail of the function of the switching sequence. Although it lacks the visual stimulation of the foregoing and, therefore, the ease of comprehension, to the trained eye it is more explicit of all detail. It takes you, step by step, through the circuitry, leaving nothing to chance:-

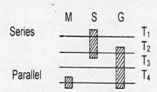

Each cross-hatched area, or "Bar", represents the site on the controller drum, or contactor unit, where contact with the specific switch is closed, energising the circuit concerned, and each transverse line represents the exact duration of closure of the particular switch. When the line is not over the bar the circuit is neutral. So, to run the eye along each line from left to right, each switch bar crossed indicates that switch is closed, or live, while crossing spaces indicates the particular switch open, or neutral. Do this, now, and you will see the same story unfold exactly as the group of drawings above.

John Markham has drawn attention to the fact that the letters M, S and G, used in Shunt Transition, were latterly often joined by a switch labelled JR as well, and that the problem with sequence diagrams in dot form was that

they did not show how the changes from one combination to the next were achieved. In the above simplest example, "S out" followed by "M&G in" gave a momentary direct line-to-earth short circuit! He concluded with the advice:

"I have always advised beware of Dot Charts. Use a Bar System instead so that the intermediate combinations are never in doubt".[2]

In the more intricate power schematics, of bridge transition, where the use of the letters M, S and G are no longer appropriate, P, J and G are often substituted instead, often with JR, for complete accuracy.

The Source of Torque of the Electric Motor or, the point where electricity is converted into rotational movement.

Torque is a twisting force, the force causing rotation: the moment of a system of forces tending to cause rotation (C.O.D.)[3] and from one's early Physics classes the "moment" is the turning effect produced by a force acting at a distance on an object. It is the product of the force and the perpendicular distance from its line of action. (C.O.D.) (That is, the piston rod of a steam locomotive acting on the crank rotates the axle, or in the electric motor the interplay of two magnetic forces – the field magnet and the magnetic force of the armature coils – that causes the armature to rotate.) In R. Brooks' "Electric Traction Handbook (control): 1954, Chapter 2, page 14, the torque of a motors is defined as the product of the armature current and the field strength.

The direction of flow of a d.c. current in a circuit may be indicated by two simple symbols – "⊙" and "⊗". Think of an imaginary dart-board dart, travelling along inside the electric wire in the same direction as the d.c. current is flowing. Then consider sections through that wire, just at the tip of the dart's point and just immediately behind the rear of the dart. The first one would look like this "⊙" – the tip of the dart pointing towards you, while the other one would look like "⊗" – the tail of the dart disappearing from you. Both indicate the direction in which the dart would be travelling and both indicate the direction of the d.c. current in the wire!

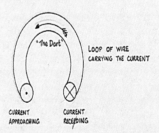

Between opposite poles of any magnets exists a magnetic force. An inert wire, static within such a field, experiences no reaction.

Stage 2 – early stages of rewinding

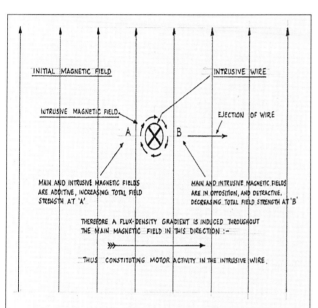

If a d.c. current was passed through such an inert wire it would immediately develop its own electro-magnetic field around itself and on one side of the wire at "A", it would be complementary to the magnetic field, while on the other side at "B" it would be antagonistic to the main field. In other words, at "A" the main, and intrusive field strengths being in the same direction would increase their total strength, while at "B" they would be opposed thus diminishing total local field strength and causing a pressure gradient towards "B" to develop.

Under the increased flux density at "A" the wire moves down the flux-density gradient from high, at "A", to low at "B", through all the main magnetic lines of force thus constituting motor activity. This action also produces e.m.f. in the wire.

It was Faraday who discovered that current would flow in an inert wire, moved mechanically across the lines of force in a magnetic field, thereby defining electrical generation. Such a wire moving across a magnetic field does not, of itself, capture energy from that field. The source of energy is the mechanical force moving the wire. No energy is extracted from the magnetic field.[4]

A wire carrying a direct current and moved normal to the magnetic field, as considered above, does experience an expulsive moment from that field which represents motoring.

Any direct current electric motor, in action, therefore experiences both of these propensities together, all the time: that is to say, while motoring, it is concomitantly generating an opposing voltage current in the reverse direction to its driving current, called the "Back e.m.f." referred to earlier. Inevitably a balancing point is reached which is the normal power output less the Back e.m.f., or from the point where the Back e.m.f is overcome.

The Back e.m.f. is ever present and, with modification, this is the precursor of regeneration.

The Approach to Regeneration:

To reiterate, the electric motor, while motoring under the application of an external electric current, is still a bundle of electric wires rotating in a magnetic field, and as such is at the same time picking up an entirely different electric current – the Back e.m.f.

This current is in the opposite direction to the applied motoring current and so opposes it, limiting it, hence the name, and the machine is both motoring and generating simultaneously.

Hirst has stated:-

"It must be realised that there is no difference between the generated e.m.f. in a generator and the Back e.m.f. in a motor, except in their directions relative to the current. Both are the same e.m.f. produced in the same way, that is by the motion of the conductors through a field" [5]

The Back e.m.f. and Balancing Speed:

The Back-e.m.f. of the motor builds up against the applied motoring voltage as the d.c. traction motor exerts both its functions of motoring and generation at the same time. As it increases in speed, it approaches, and ultimately reaches a state of equilibrium between the two, called the Balancing Speed. It then settles down to run steadily at this value without further active control by the motorman, at a steady speed. It can therefore be defined as the top speed reached by a vehicle, or train, on a level tangent track.

When an electric train, or tramcar, running under load on the level, should reach an incline, the speed naturally falls automatically and along with it, its opposing Back e.m.f. The two reach a new equilibrium with each other commensurate with the state of gradient and load together with head-winds and cross-winds. The motors automatically reduce speed and an increase of current occurs leading to an increase in tractive effort. This all takes place without any necessary adjustment of the controls by the motorman and the train or tramcar continues climbing the up-grade competently – an immensely useful characteristic of the d.c. traction motor. Tractive effort is dependent on current, although not in an absolutely linear relationship, when using series-wound motors.

The Back e.m.f. in Motor Protection:

The only available opposition to the fully applied voltage is the very low resistance value of the motor's windings. Starting resistors require to be switched into circuit at No. 1 notch to counter this. However, upon rotational speed increase, the Back e.m.f. obtrudes, offering further motor protection by countering the applied voltage as well as increasing speed

and allowing progression to the second series notch. This process continues, notch-by-notch, in series, until the opposing Back e.m.f. approaches 80% of the applied voltage and the resistors can be cut-out altogether at full series notch.

Series Control:

The nominal line voltage, or electrical pressure, is regulated at substation level. Initially, in Glasgow, for instance, this was 500 Volts. Much later on, with increasingly high demands with more, and heavier cars carrying increasingly heavy loads of passengers – especially at rush hours, it was increased to 575 Volts. The London Underground also commenced at 500 Volts, rising to 630 Volts to cope with the same demand while on parts of the national network, especially Richmond, its trains get 750 Volts. [6]

Tramcars starting from rest cannot cope with 575 Volts shot through their machinery and this was reduced by the use of resistors before reaching the motors, burning up the excess voltage in heat, there being initially virtually no opposition to the relatively heavy voltage. The effect was augmented by putting the motors into series circuitry as previously described. The car would begin to move slowly concurrently with the resistors heating up rapidly in opposition to the relatively heavy line voltage.

As the car began to move, gradually the Back e.m.f. began to develop adding further opposition to the motoring current, and so to the further protective effect to the motors until, with the increase to half-speed at full series, the Back e.m.f. has reached the point where all resistors can be cut out and the motors can comfortably cope with half line voltage.

First Series:

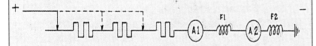

In first series the full line voltage is diminished by the full bank of resistance (the resistor) each motor taking half of the residue.

Full Series:

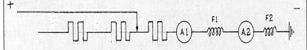

As the motorman advances the main controller handle, notch-by-notch, sections of the resistor are cut out, adding increment after increment of voltage to the motors in a gradual manner so as not to overload them as the car picks up speed. These intermediary steps, not illustrated here, are readily conceived.

Starting in series has the advantage of both the starting losses in the resistor being minimised together with economy of the low-speed running notch in full series for the negotiation of traffic snarls in city centres. Had both car motors been started from rest in parallel, resistor loading would have been excessive, carrying the full accelerating current of both motors with high ohmic losses. [7]

Transition:

Originally, and for a long time, there was only the one method of transition from the series state to the parallel state of series-wound motor operation. This involved the isolation of the No. 2 motor for a transient moment as the power circuit was rearranged for parallel working. This threw the whole weight of the maintained running of the tramcar on to the No. 1 motor, momentarily, with the latter machine carrying the full line voltage, alone, until re-adjustment brought in the No. 2 motor. This must have put great strain on the No. 1 motor for the miniscule period of 0.4 sec. This period was called the Short Circuit Transient of the d.c. series motor and has been worked out for the electric railway motor which differs little from its tramway cousin.

Bridge transition, on the other hand, came along a good deal later, and was a tremendous advance on the short circuit transition in that at no time was there need for the complete cutting-out of a motor, both remaining in circuit throughout. This circuitry was ideal for four-motor, double-bogie tramcars and most useful even in modern two-motor vehicles. It was more expensive to install and to maintain, requiring two separate sets of resistors – one for each single motor (or pair, depending on the type of tramcar).

Short-Circuit Transition and Short-Circuit Transient:

First, from the characteristics of the series-wound motor, the process of completely cutting-off a motor across both fields and armature while in continued running, causes – as its name implies – the short-circuit transient of the d.c. series motor, occupying an instant of 0.4 of a second, in its entirety. During this extremely short period of time a series of electro-magnetic events occurs within the now isolated, but mechanically gyrating motor. Instantly, the Back e.m.f. builds up rapidly, its opposing, motoring current having been withdrawn into a "Reverse Current", or "Generating Current" of considerable amplitude, but of very low voltage.

As this "Reverse Current" rises it rapidly demagnetises the motor's "Residual Magnetism", being in reverse direction to the withdrawn motoring current, so rendering the motor momentarily electrically inert: the "Reverse Current" having slumped to zero. [8]

During this infinitesimal period the reversed polarity of the "Reverse Current" demagnetised the motor's field to zero polarity, thus preparing the situation for the re-establishment of positive polarity for parallel working – all this within 0.4 of a second!

Stage 4 – nearly complete

The four actual stages in Transition
- (a)   Reinsertion of series resistance
- (b)   Short-circuit of No.2 motor, (when it demagnetises)
- (c)   Open-circuit of No.2 motor
- (d)   Connect No.2 motor in parallel with No.1

There can be a fifth:-
- (e)   Reducing the series resistance[9]

Pictorially these steps may be drawn as follows

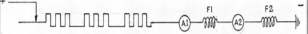

The first step in Shunt Transition involves the reinstatement of the full resistor into circuit with both motors still in series, as above.

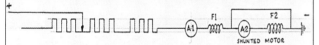

The second step shows reduction of resistor, together with shunting of No.2 motor, while still in circuit

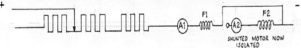

The third step in transition shows the complete isolation of No.2 motor, still under substantial resistance. [10]

If, however, the connections of either armature or field are reversed within this brief, Short-Circuit Transition period, and before the Back e.m.f. has crashed to zero the motor will then re-build up and continue to deliver a very heavy current round the short-circuit. This will be in a direction contrary to the motoring current and this current is what is used for rheostatic braking.

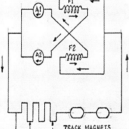

The rheostatic braking circuit most commonly used in this country, showing reversal of field connections and their crossing with the armatures of the opposite motor. [11]   There were others.   Transition in regenerative braking was more intricate and will be tackled later.

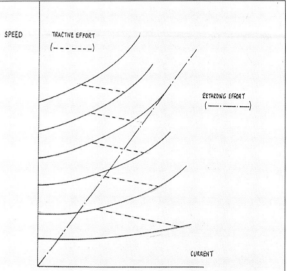

Maximum voltage in rheostatic braking can be up to about 2.5 times the maximum motoring voltage on electric railway trains.   On tramways this figure could reach about 1.75 to 2.0 times the motoring voltage. [12]

Harking back to the third step in shunt transition and still under substantial, if not full resistance, the short-circuited No. 2 motor is realigned in parallel with its partner.   Each is now taking the full line voltage – less resistor and circuit losses and they have achieved First Parallel Notch, as below:-

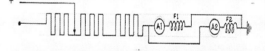

First Parallel Notch

From first parallel notch to full parallel notch, step-by-step resistance segments are withdrawn from circuit allowing incremental increase in voltage across the motors just as occurred in the progressive series control.   The intervening steps to full parallel control are readily conceived and are not illustrated here.

To round up on Shunt Transition, it was outstandingly the most common approach to transition in Britain throughout the whole of the first phase of tramway development from the early 1890s until the early 1960s.   The pros and cons can be summed up thus: the process is simple, rugged, less expensive to install and maintain and uses fewer expensive contactor switches than Bridge Transition.   The drawbacks lie in the fact that for a moment or so one of the motors is out of action and the full weight of the car is carried by a single motor putting repeated heavy strain on to it, despite still working under substantial resistor protection.   Also, during that split-second of disconnection, the car is still moving, the now isolated

Please see the full description for this picture in its enlarged form on page 188.

motor's armature is being mechanically rotated and so acting as a generator using the residual mechanism of is fields. This, however, so rapidly drops as to be of little consequence. Although momentary before the establishment of parallel circuitry and the withdrawal of resistors, there are disadvantages in the obvious reduction of torque and diminution of overall braking effect.

Bridge Transition:

Bridge transition was little used in association with the direct manual controllers of the same phase of tramway usage. Automatic control allows the conditions at the moment of transition to be so explicitly determined that the value of current undergoing interruption is always at a normal level. All motors are continually under power with none being cut-out as with Shunt Transition, so that there is no loss of power or of braking mechanism. Another feature of Bridge Transition is that each motor (or pair of motors in a modern double-bogie car) requires its own separate set of resistors, adding to both weight and expense.

The Bridge Transition Schematic:

FIRST SERIES

SECOND SERIES

FULL SERVICES

TRANSITION

FIRST PARALLEL

FULL PARALLEL

This is, perhaps, the most simplified exposition of Bridge Transition. [13] Step-by-step the rearrangement of connections may be seen at a glance. Its outstanding merit is that of uninterrupted excitation of both motors throughout the whole process and is easily seen. Transition is effected in a single step but the resistors must be divided into two separate sets of exactly the same characteristics to allow each motor, in parallel, to have its own resistor unit connected in series with it. At full series point the two motors are directly connected in series, with all resistors open. At the transition step the two sets of resistors are connected across the supply, respectively, in the manner of a Wheatstone Bridge.

MOTOR No.1   F1   RESISTOR No.1

A1   EQ

EQUALISER SWITCH

RESISTOR No.2   F2

A2   MOTOR No.2

The Wheatstone Bridge

As transition occurs the bridge connection equalises the flow of current between the two parallel paths of each motor and the opening of the switch "Eq" completes the traditional parallel combinations of the two motors. Thus bridge transition allows the normal accelerating torque of both motors to continue uninterruptedly throughout the entire accelerating phase without a break.

When we approach regenerative braking, bridge transition technology is found to be far too complicated for economical tramcar usage and is reserved for simple series-parallel control with four motors, where it was obligatory, although it had been tried on experimental two-motor equipment.

For further notes, see the end of this chapter.

Parallel Control:

In the series circuit, excitation passes through both motors in series, each motor absorbing approximately half of the available line voltage, less resistor and circuit losses. As has been seen, at full series notch the car is working nominally at half speed, depending on a lot of variables.

Series

+   A1   F1   A2   F2   —

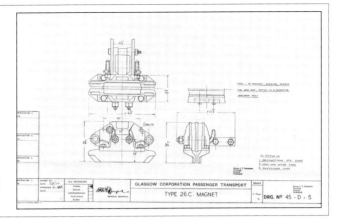

A reproduction of an original Glasgow Corporation Transport drawing bearing the signature of General Manager ERL Fitzpayne.

In the parallel circuit, however, each motor is realigned across the full available voltage (always less resistance losses) and as such they are now subjected to virtually full line voltage, each giving approximately full speed, in the simple traction circuit.

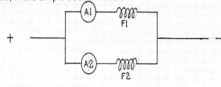

Parallel

So, the simple direct hand-operated controller, of old, gave the car only two speeds forward and for economy the motorman took his car up to top parallel and returned the controller handle smartly to the "OFF" position once adequate way was on, then coasted to his next stop.

Parallel circuitry is achieved on the last step of transition in which both motors are found in parallel with each other and in series with heavy resistors, each taking potentially full line voltage.

The steps to full parallel are simply the sequential withdrawal from circuit of step after step of resistance until the two motors are set across the full supply voltage of the overhead, uninterrupted by resistors, at the top parallel notch.

This is then the car travelling at its nominal full speed, on the level, and cannot go any faster unless on a falling gradient or, unless the traction fields are shunted to diminish their field strength and thereby also diminish their Back e.m.f. This, in turn, allows greater exploitation of the line voltage in a further increment or so of speed. The principle of "Weak Field Control" is usually arranged to give extra increments of "economical" higher speed. This is studied in further depth in the Appendices.

Electric Braking:

By the mid-1950s more or less from its inception and during the first general development of the tramcar there existed two methods of electric braking –perhaps three if regenerative braking is included. These were "plugging" and Rheostatic braking, of which the latter outshone the former. Regenerative braking never quite made the grade until very much later on!

Plugging:

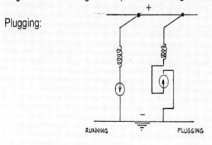

In "Plugging", by reversing the current through the armatures the torque of the motors is reversed exerting a reversing tendency within the motors and therefore a braking effect on the progress of the car. In the running position above, the car is in balancing speed and the Back e.m.f.is virtually equal to the overhead supply voltage, and opposing it, so that only a small voltage is available to drive the normal current through the small resistance of the motors. In the plugging position the Back e.m.f. is in the same direction as the supply so that at the moment of switching, twice supply voltage (often over 1000 Volts) is available and an enormous rush of current would occur. Limiting resistors must therefore be inserted in series with the motors. In the process of braking, therefore, supply volts plus kinetic energy of the moving car require dissipation in resistors in a very energy-wasteful process despite the efficiency of the braking. [14]

Plug braking may be used at the point at which the rheostatic braking fades away and the car has not yet come to a complete rest. It is quite possible actually to arrest the car without assistance from mechanical means. The technique is to return the controller handle to the "OFF" position then put the reversing handle/key into reverse and apply one notch of power. This is often less severe than the Emergency Brake providing the sped is low. But the car will set-off backwards after it comes to a halt if the controller handle is not returned to "OFF" as soon as it becomes stationary. The technique is the first part of the "Last Resort Brake" and possible on all two-motor cars having transition. [15]

Rheostatic Braking:

We have already taken a look at how this is effected. The motors are disconnected from the supply into a closed short-circuit pattern, but including variable resistors. The kinetic energy of the moving car, in rotating the motors, causes them to generate, which current experiences graded dissipation in the resistors and track-brake shoes. Being a closed circuit without overhead contact, rheostatic braking continues to work even in the face of a dewirement, or derailment, or both! The motors, however, must be in parallel for braking purposes with both machines equally sharing the load to prevent large currents building-up in the opposite direction, short-circuiting each other. No two machines to the same specification are exactly equal in performance, therefore the stronger would overpower the weaker one into motoring instead of generating. This situation is controlled by crossing the fields and armatures of each with the other such that the armature current of the more powerful machine strengthens the field of the weaker, resulting in equal excitation of both. [16]

In the rheostatic brake, operation is by the movement of the controller handle in the opposite direction. The first brake notch closes the motor circuit through the full resistance and inserts the graduated resistance. Further movement of the main controller handle, backwards, gradually cuts-out the resistance which, at the last notch, is entirely eliminated. This

151

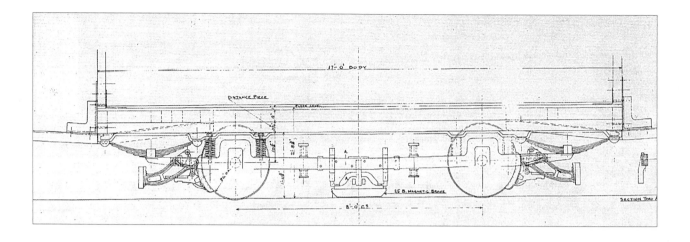

graded brake control eliminates the tendency to skid the wheels at

ordinary speeds and to a large extent eliminates the severity of braking application. However, constant use of the rheostatic brake builds motor heat to a high temperature and tends to shorten the life of the insulation while increasing the tendency to flash-over at the commutators at sudden application.

Its advantages were:
    Low first cost
    Rapid retardation
    Simplicity of adjustment, and mechanism
    Uses the car's kinetic energy
    and produces its braking effect from armature reaction

The disadvantages were:
    The need for necessary skill to apply it efficiently
    The tendency to skid the wheels – again depending on the skill of the motorman
    The braking effect depending upon continued revolution of the wheels
    The tendency to increase the temperature of the motors – and their deterioration
    The large number of contact points in the electric circuit

The rapidity of application and the power elicited varied with the speed of the car and at some minimum speed the brakes would not come on at all due to the motors failing to build-up their fields. However, the rheostatic brake is superior to other forms of wheel-brake in that the retardation did not fall off with increase of speed as they used the coefficient of wheel adhesion, which is dependent on speed. This advantage, however, would vanish if flash-over-speed is reached. It can also be a useful run-back preventer if so fitted. It is a very powerful brake in skilful hands but as a service brake it puts a heavy strain on the electrical equipment.[17]

Rheostatic braking is defined as when the electrical current generated by the motors in short-circuit is dissipated as heat in the vehicle's resistors.

It is the method of choice for bringing a tramcar to a stop but not for controlling the speed of a heavily laden car – or train – on a long descending grade. In that case the regenerative braking system would offer the most suitable choice.

Mr Gerrard, Assistant Electrical Engineer to Glasgow Corporation Tramways Committee, in the opening decade of the twentieth century, made extensive tests on electric brakes, in all sorts of weather and track conditions. This led to his being much in demand in talking on the subject. His diagram of connections for rheostatic braking obviously employed a voltage equaliser connection between the two parallel motor circuits to prevent motor-generator effects around the closed circuit between the two

machines. His original circuit diagram suggests the use of the earthing track tram-rail as a component of the closed rheostatic circuit. This arrangement would have been contra-productive in that a derailment accident would have rendered the rheostatic brake inoperable through braking of the rheostatic circuit.

The Gerrard Rheostatic Brake, braking schematic. The generated current was fed backwards through the resistor, or rheostat, in those days, to ground, so offering powerful deceleration.

The Back e.m.f. is not the prime mover in rheostatic braking. Once the main controller handle has been returned to "OFF", on trams with rheostatic braking, the motors are electrically "dead" and unless the tram rolls backwards with the reverser key in the forward position, the Back e.m.f. is flowing the wrong way! An anti-clockwise movement of the main handle on the typical controller from the "OFF" position causes a Power-Brake changeover barrel to throw. This effectively reverses the polarity of the armature connections and, once the lines of force in the magnetic field (its residual magnetism) are cut by the conductors in the armature, the small voltage is fed through the field coils. These become excited, increasing the lines of force and the power escalates. In tramway engineering it is recognised best practice to leave the field-coils permanently connected in the same polarity, and to do the switching in the armature connections. In this way the residual magnetism in the motor frames is not weakened by repeated reversals.

"Apart from the marked economy of deriving free braking current from the car's motors, the advantages of the rheostatic brake system also include the braking system being totally self-contained on board the tram, therefore it is unaffected by dewirement or derailment. Even if the brake notches in the controller become jammed, the brave motorman can still achieve a total application of the electric brake by first switching off the main circuit breaker, then pulling the reverser key into "reverse", and hitting any parallel notch with the main controller handle. Doing this reverses the armature connections, and sets-up the parallel connections of both motors – thus within the confines of this "box" in the circuitry, an effective short-circuit is formed across the output of the generating motors. Although a series-wound motor acting as a generator has a characteristic whereby the greater the current flowing through the field windings, the greater is the voltage of the output, this has no opportunity to "blow" the bulbs in traction voltage lighting. The only damage that the generated high voltage (0ver 1000 volts when an emergency braking application is made) of the braking current can inflict is within the traction system's motors and controllers. Drivers are taught never to bring the main controller handle back to "OFF" from a heavy braking application until the tram stops. This is the Main Emergency Brake routine, when everything else fails! It is also referred to as "Last-Resort Braking".

The main disadvantages of rheostatic braking are the waste of energy, in that power is converted into heat in the starting resistors being employed as a load bank during rheostatic braking; and the minimum half-second time-lapse necessary for the braking circuits to become energised (only when this is done can the braking effect commence). The latter problem is dealt with in post-1950s German designs in that the 28 Volt battery supply used for auxiliary equipment (doors, ticket cancellations, etc.,) is injected into the motor field connections on the first braking notch. The time delay is avoided by this means, which "excites" the field windings in the braking mode. The old system needs time for the residual magnetism of the steel motor frames to begin producing a small voltage. This is then fed into the field windings, providing many more lines of force, a higher voltage is produced and the rheostatic brake then really begins to "bite". The inexperienced motorman usually has, by this time, panicked and moved the controller handle to a higher brake notch than appropriate, which, when the brake becomes effective, results in a severe lurch for the passengers! Cases are recorded when trams released from overhaul and newly fitted with motors which have been dormant in store for many months, and so lost their residual magnetism, being unable to produce any rheostatic braking effect on their delivery journeys! The other dangerous shortcoming is that burnt or dirty contacts inside the controller can seriously inhibit circulation of the initial low voltage, and prevent the all-essential build-up to energise the field coils – thus leading to brake failure". [18]

ast-resort Braking:

First consider motors in parallel:-

Then in coasting phase:-

Say $V_1 > V_2$. This will cause current to flow anti-clockwise in the loop. Therefore Motor No. 2 will be motoring and Motor No. 1 will become demagnetised and therefore no current will continue to circulate and the motors will stop.

Last-Resort Braking involves throwing the reverser, if coasting, but if in the motoring phase first the circuit-breaker must be thrown off, then the controller handle put into any parallel notch, then the reverser is thrown.

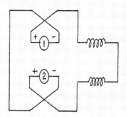

Again, say $V_1 > V_2$. This causes No. 1 motor to self-excite, and demagnetises No. 2 motor.

Then motor No. 1 continues to self-excite and increases its own voltage, thus causing more current to flow.

Motor No. 2 then suffers excitation in the opposite direction:-

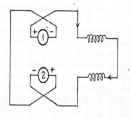

Motor No. 2 starts to generate, and self-excite. Both motors now in series ring, and on a short-circuit.

This circuit is inherent as a "run-back" brake in cars fitted with cross-field braking rather than bus-field. [19]

Rheostatic and Regenerative Braking principles are basically similar, varying mainly in the disposal of regenerated, or reversed current produced. In the rheostatic type, this current is dissipated in the vehicle's resistors, while in the regenerative variety the current – boosted by the increased power of the compound winding of the motor's fields – is

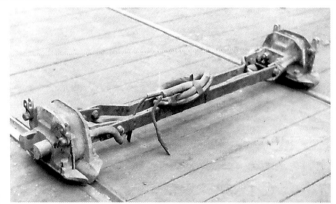

Metropolitan-Vickers Type 26C track Magnet unit.

The field coil separating the N and S magnetic plates of the track magnet may be easily seen, as also the energising cables. The external pole-plates had a pair of very obvious D-shaped studs which loosely embraced a strong steel guide bar, the downwards extension of the brake suspension bracket, bolted to the main side frame of the truck. The upwards extension of this guide-bar was bent inwards below the car body sills and branched fore-and-aft to provide two suspension brackets from which powerful retaining springs descended to attach the lugs on either top shoulder of the pole plates. These springs held the magnetic brake pole plates free above the rail-heads, yet retained them to position accurately above the respective rail-heads on either side of the truck allowing the magnets to clamp onto the rails in action, by powerful magnetic attraction.

returned to the overhead. In both, the braking effect diminishes with reduction of speed and ceases altogether <u>before</u> the motors are brought to rest.

They differ in that, the rheostatic system, the braking effect is unaffected by derailment, or de-wiring – being a completely self-contained system within the tramcar, and therefore continues to function regardless of such possible incidents. The regenerative braking system, however, is utterly dependent upon the integrity of the overhead supply, and rail return. If either failed, the braking effect is broken and totally lost. For this reason, the automatic linking with the compressed air system of braking is imperative.

Some General Considerations about Braking:

Service stops should be, while fairly regular, considerate of city travellers' interests in shopping or going short-distance business messages. Short distance travel should be provided at very reasonable prices to encourage maximum use of the trams to save time in preference to walking. Journeys like these have always brought in the backbone of revenue of city tramways. During the first phase of British tramways, city fares rarely exceeded 2½ to 3d (1p). Today the cheapest fare on typical modern tramways is £1·00. Of such is spelled overwhelming loss of earnings. Sadly the blame is political in statutory inter-stop distances!

The shorter this distance is, the greater the acceleration, the higher maximum speed between stops and the increase of deceleration for the next stop, irrespective of intermediate braking and acceleration due to congestion of traffic. All is necessary to maintain overall schedule speed. This, naturally, brings into consideration the limit of retardation that passengers (particularly standing, or moving) can comfortably tolerate, and vice-versa in terms of acceleration. It has been estimated that at 3 feet/second, per second is about the upper limit of comfort, either way. [20]

The control of speed on downhill grades exceeds the normal, level, requirements by the braking effect necessary to neutralise the acceleration due to gravity. On a steep incline this is a figure far in excess of that on the level. The kinetic energy of a moving car is proportional to the square of its speed, on a straight line, while the tendency to overturn with a double-decked car at curves, due to centrifugal force was also reckoned proportional to the square of the speed. [21] Even on the level, at 10 m.p.h., a car has only half the kinetic energy it would have at 14.14 m.p.h., and only a quarter of what it would have at 20 m.p.h. It is this energy that has to be absorbed by the brakes, of any sort. It has been estimated that in assessing the danger of any gradient, of any length, the total fall multiplied by the gradient percentage gives a general figure, regardless of curves, density of traffic, crossroads and so forth, for an assessment of the relative danger it represents. These ignore other curves of loss of power such as greasy rails, wet leaves on the track, the development of flats on the tyres

and skidding due to locking of the wheels in heavy braking. Pitching and rolling of the vehicle also diminished the braking effect due to the variation of the distributed car weight on the braking wheels and this was enhanced on double-decked trams.

These few asides are only a glance at an important aspect of street transport organisation and very little, if ever, is this considered outwith professional interest. They add great weight to the process of braking and to the types of braking employed.

The Genesis of the Rheostatic Current

We have come across "Back e.m.f." already. It is simply the reversed current of an actively motoring motor and only occurs in the presence of active motoring. It is a generic factor of it – hence its name and positive relationship to the motoring current. No motoring current – no Back e.m.f! When the motoring current is withdrawn the Back e.m.f. ceases, but a reverse direction current immediately takes its place and continues to flow. This is a new variety of reverse current, unrelated to a motoring current, and may not therefore be referred to as a "Back e.m.f." It is a new entity, a reversed current of different origin, related to the residual magnetism in the motor's field, and should be called the "Reversed Current", or more conveniently, the "Generated Current". This is seemingly a quibble in words, perhaps but functionally not so, even though both are travelling in the same direction and are apparently continuous and without obvious alteration – except in power. It is the Rheostatic Braking current.

The steps in auto-generation of the Rheostatic Braking current may be visualised as follows:-

1. Normal motoring, productive of its reactive "Back e.m.f."

The motor is now disconnected from the supply voltage and reconnected in isolated circuit with a resistance and the brake magnets. There is no further controller influence of the isolated brake short-circuit other than to control the variable resistance application. The circuit lines are now representative only of the direction of the isolated generated current.

2. Motoring current is cut-off. Therefore armature polarity and Back e.m.f. are destroyed. The car is still rolling under its own momentum. Field polarity is maintained by reluctance and now produces a new reversed current.

3. Residual field magnetism continues to produce a Reversed Current causing reversal of polarity of both armature and field, neutralising the Field Magnetism and preventing generation.

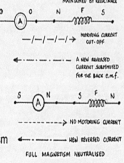

A close-up view of the magnet looking from inside the truck. Both photographs were taken by Struan Robertson on one of his many visits to Coplawhill during the course of his researches.

The Short-circuit Transient: If a running series motor, part of a series circuit, is short circuited across armature and field, it first generates a reverse current. However, since the direction of this current through the field opposes the generated voltage, the machine is rapidly demagnetised, becoming electrically inert. The magnitude of the reverse current is considerable but as the voltage in the short-cut is very low, the amount of energy in the circuit is limited and the duration of the transient condition very short – less than 0.4 second. [22]

4. The new Reversed Current enhances the residual field magnetism and the motor builds up as a generator, producing the Rheostatic Braking current.

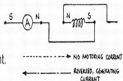

The reversal can theoretically be of either the field (as above) or the armature leads. It was customary in tramway practice to reverse the armature leads in order to save the field magnets from repeated reversal of polarity.

The reversed current very rapidly demagnetises the motor's fields which become electrically inert. The amperage was considerable but the pressure (voltage) was low, rendering the amount of energy limited and the duration of the transient connection very short. Connecting either the field, or armature, in reversal while within this short-circuit transit period allowed the machine to build-up and to continue to deliver a heavier current round the rheostatic braking circuit.

Because of this heavier current, the motors were put in parallel for braking, so it was essential that the machines shared the load equally otherwise that which built-up first would over-excite the other, causing it to build-up in the opposite direction, thus short-circuiting one another. There were two main methods of avoiding this by equating the load between the two motors, first by the use of an equalising bar in the circuitry, and the other by cross-connecting the fields and armatures. This made the more powerful motor excite the less powerfully excited machine, thus equalising excitation of both. [23]

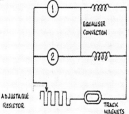

The avoidance of short-circuiting in parallel operation using the equaliser connection. For the cross-fields and armatures, see previously. [24]

But how very close to Regenerative Braking

In order to use this same "Reverse Current" for Regenerative Braking, it was necessary artificially to increase it to a power well above the line voltage for it to be accepted by the overhead. This explains the need for special compound windings in the regenerative motor fields. The resultant enhanced voltage also acts to stabilise the effect of the series-wound field coils, the higher the voltage produced. So, when a powerful extraneous exciting current was fed through these special compound windings of the fields of the dying motors, their field strength revived and the armatures burst into life as powerful generators. To reach that state of affairs in tramway practice it is necessary to cross the fields and armatures between the motors of the car. Crossing of the field and armature leads is only done for reversing, or in rheostatic braking. [25]

REFERENCES: CHAPTER 10

1. Markham, J.D.: personal communication: 23 8 02
2. Ibid
3. Concise Oxford Dictionary
4. Ibid (Markham): 23 8 02
5. Hirst, A.W.: Electricity & Magnetism for Engineering Students: 2nd Edition, 1947: Glasgow: ch XV:16:363
6. Ibid (Markham): 23 8 02
7. Brooks, R.: Electric Traction Handbook (Control): 1954: ch III: 19
8. Ibid (Brooks, R) : 1954
9. Ibid (Markham): 23 8 02
10. Everett, K.E. (Ed): Lett's Electrical Diary: 1961: 161
11. Ibid (Brooks, R) : 1954
12. Markham, J.D.: personal communication: 7 9 00
13. Ibid (Everett, K.E.): 161
14. Starr, A.T.: Generation, Transmission & Utilisation of Electrical Power: 1949 – 2nd Edition: ch XII- "Electric Traction": 377-8
14. Starr, A.T.: Generation, Transmission & Utilisation of Electrical Power: 1949 – 2nd Edition: ch XII- "Electric Traction": 377-8
15. Ibid (Markham): 21 10 02
16. Ibid (Starr): XII: 377-8
17. Tramway & Light Railway Association: Report of Special Committee on Braking Arrangements & Sanding Gear on Tramways: T&L.R. Assoc. Special Brakes Number: 1908-September: No.69: pp 816-8
18. Tudor, The Reverend D.: personal communication: 22 11 01
19. Ibid (Markham): 7 9 02
20. Ibid (T&L.R. Assoc): 782
21. Ibid (T&L.R. Assoc): 783
22. Ibid (Brooks) : 20
23. Ibid (Starr): XII: 378
24. Ibid.
25. Ibid (Markham): 22 9 03

THE ELECTRICAL SUPPLY

John Markham takes us through the processes of electrical supply to tramways and the impact regenerative tramcars had on the infrastructure.

JUST IN TIME

To consumers such as us in a domestic situation, electricity is a source of power that is highly adaptable in its application. Present day households accept this without a thought and take it for granted that it will always be there and really only notice when, for some unknown reason, the supply fails. Very few people realise, however, that their supply is received from an organisation that has no ordering system nor can it hold stock of the commodity it is selling. It is the epitome of 'just in time' manufacture which industry generally started to introduce during the final quarter of the 20th century.

Electricity is not a prime source of energy. It is an energy distribution system. Some prime source has to drive an energy converter (generator, alternator, fuel cell, solar cell, etc.). The actual source can take many forms: coal, gas, oil, nuclear, water flow (hydro or tidal), wind, solar, etc., depending upon local circumstances. Coal-fired steam-operated stations were by far the most numerous during the traditional tramway era.

The principles that follow are not concerned with the prime source but with the energy distribution system, which takes the next stage in the supply chain.

That electrical machines are reversible in terms of power flow has been amply demonstrated in other chapters but for all practical purposes, prime energy sources are not. The chief benefit of tramway regeneration lies in the ability of the supply and distribution to find another load which is able to absorb the regenerated power on offer elsewhere on the system. Interconnection is the key to this but this must be developed with care to ensure that the results in all circumstances keep the network parameters, such as voltage (and frequency, if ac distribution is involved) within defined limits.

The early regenerative tramcar trials – and sales – involved small numbers, often only one, to operate within an existing fleet and working on tram systems small enough to be fed from one power station. Providing the regenerative car operated only during the core part of the day, there was a high probability that somewhere on the system there would be other trams taking power and thus capable of absorbing that returned to the overhead line by the sole car when generating. This would still be true if the electrical connection between the two was via the power station traction busbars. The power flow could easily be across town.

The situation changes as the proportion of cars in a fleet capable of regeneration increases. In these circumstances there is an ever-increasing risk that more regenerated power is on offer than can be absorbed by the total power demand of other trams which are taking current at the same time.

PROBLEMS IN RAWTENSTALL

Nowhere was this problem more fundamental to the power system than at Rawtenstall which, from its inauguration in 1909, had a tramcar fleet of 16 and was 100% regenerative. The Raworth equipment installed on the trams used regeneration as a prime braking method and this had to be taken fully into account in the design of the power station. Having the same man, CLE Stewart as borough electrical engineer and engineer to the tramways eliminated interfacing difficulties between traction and the supply to other customers.

The construction of Rawtenstall Corporation's power station for domestic lighting and traction was contemporary with the construction of the tramways for electric working. The steam generator

The chief benefit of tramway regeneration lies in the ability of the supply and distribution to find another load which is able to absorb the regenerated power on offer elsewhere on the system

sets installed produced dc power at either 460 volts for lighting or 525-550 volts for traction. The lighting supply was made 3-wire by the use of a balancing set whose mid-point third wire would be earthed. Domestic consumers would thus receive a 230 volt dc supply with either earthed positive or earthed negative. 460 volts was available for light industrial use. The traction supply was earthed negative by virtue of being connected to the rails forming the return circuit. Direct connection between lighting and traction was thus impossible because of this fundamental difference in their earthing systems, even though every generator could be used for either lighting or traction purposes.

Rawtenstall could, however, transfer energy between the two electrical systems, thus achieving a good balance of load between the various generators in use, and, importantly for the tramway, transfer surplus regenerated power over to the lighting system from which it was resold to somebody else.

How was this achieved? Answer – by two rotary converters that ran in an inverting mode to produce 50 cycles per second three-phase ac. This, after transforming to 3000 volts, was exported to local substations in Crawshawbooth and Whitewell Bottom for local use. Additionally, the two converters, when mechanically coupled together by a clutch, could act as a dc motor-generator between traction and lighting while simultaneously providing ac. The power station also had a battery of lead acid accumulators capable of supporting the lighting overnight when the steam plant was shut down, and acted as a load buffer at other times.

Failure to make provision for power absorption created numerous problems with some early systems and catastrophic failure of generating equipment for the Yardley tram route of the City of Birmingham Tramways Company was a classic example. The same philosophy had to apply to towns where regenerative trolleybuses replaced non-regenerative tramcars. Hastings and Wolverhampton are examples. Perhaps the reader will forgive a digression at this stage.

Failure to make provision for power absorption created numerous problems with some early systems and catastrophic failure of generating equipment

THE HASTINGS COUPLER

Hastings was something of an innovator in the tramway and trolleybus era. The development of the Hastings trolley wheel was one, though its appreciation is not appropriate for this text. Another, whose background I heard from an observer at the time, is of relevance to the regen. story.

The Hastings system, like many others, was dominantly radial in character. One substation fed the whole of the town centre but there were other isolated and independent smaller substations

The Parade and Pier at St Leonards-on-Sea with Hastings tramcars 42 and 46. Note the trolleys are tied down while operating on surface contact at this location.
(Photo: BJ Cross collection)

Hastings tramcar 31 passing beneath the Alexandra Bridge.
(Photo: BJ Cross collection)

158

in the suburbs. The power supply was provided by an organisation completely independent of the Tramway Company, the latter buying its electrical power for its trolleybuses from the former. While the regenerative equipment served to minimise the energy consumed, the effect of traffic bunching in the town centre created significant peaks of power requirements which had an unacceptable effect on maximum demand. For this their electricity supplier had a punitive tariff. The manager and engineer of the Hastings Tramways put their heads together and determined that the maximum demand would be reduced if the town centre load could be spread over several substations. They devised a scheme for their suburban substations to feed sufficient power into the town centre to enable its own substation maximum demand to be reduced, but without adversely affecting their service operation in any way.

For the next part of the story to be possible, the town centre substation must, by this time, have been converted from machines to mercury arc rectifiers otherwise unwanted circulating currents could become established.

An automatic switching device was installed in the section boxes associated with the section insulators normally separating the substations to enable them to be coupled together. They also set the suburban substations to a higher voltage than that in the town centre so, at light loads, the town centre substation lay dormant and made no contribution to the traction supply.

The scheme worked. Initial town centre load was shared by the outlying substations but as it increased, resulting in voltage drop due to the distances from the supplying substations, the voltage in the town centre became depressed below the threshold of the town centre substation and it immediately began to contribute. The result was a reduction in the maximum demand on the power company by the town centre substation, an insignificant increase in maximum demand by the suburban substations and a lower overall charge for the same total amount of power consumed.

The automatic switching device became known as the Hastings Coupler. There was another benefit. Having interconnected the substations in this way the whole system became one network for regeneration purposes.

Hastings couplers became widely used by towns adopting regenerative trolleybuses. Manchester, as an example, employed them throughout its trolleybus network from their inception in 1938. Manchester's General Manager, RS Pilcher, had insisted that his system should employ the best technology available. But we are jumping ahead too far in the story.

OVERSPEEDING ARMATURES

We have seen how Rawtenstall took full cognisance of the need to make the supply system capable of absorbing regenerated energy and how, with only a very small percentage of regenerative capability, supply systems could usually cope. There were, however, serious limits to this. It depended very much on the type of dc supply machinery used. It is obvious that a steam-driven dc generator cannot accept regenerated energy but this was also true for some types of conversion plant, notably the rotary converter – much used in tramway traction supplies. The rotary converter is, fundamentally, a machine capable of accommodating reverse power flow. However, those for tramway use, in addition to the shunt-field winding necessary for maintaining the machine in synchronism with its ac supply, were very often fitted with a series field winding. The series field was used to provide a degree of compounding to help keep a reasonably constant voltage on the trolley wire under wide variations of load, and to lock the armature to synchronous speed under heavy, and overload conditions. This was fine as long as the dc current did not reverse. We have seen how, when regenerating, the series field on the traction motors decompounds and opposes the shunt field, so it will be understood that reverse current flow through the series field of a rotary converter will also act in a decompounding manner. A little bit (depending on the size of the machine) is not a disaster but the time arises when the reduction in total field flux is such that synchronisation with the a.c. mains frequency cannot be sustained. When this occurs power transfer back to the ac side ceases and all the regenerated energy goes into accelerating the armature of the rotary converter. Unless this is detected immediately by an overspeed device and the ac and dc power connections are broken, disintegration of the armature becomes a high probability with any damage arising from commutator flashover being additional.

It was not impossible for the polarity of generators at power stations to become reversed. It could, for example, happen if only one relatively small capacity generator was feeding the busbars. A heavy reverse current through the series winding of the compound field could overcome the machine's own shunt field and residual magnetism, reversing the magnet's polarity of the machine. It would then generate wrong polarity electrically, so inverting the shunt field excitation previously produced through reverse current in the series field. The generator and regenerating tramcar would then be in series with each other and feeding into a ring short circuit. One could expect damage with some violence under these circumstances as the combined energy of the prime mover and the regenerating tramcar are fed into the same short circuit fault.

One could expect damage with some violence under these circumstances

If there was no series-compounding field on the generator this particular hazard would not occur but, under regen. conditions, the generator could become a motor driving in the same direction as the prime mover with a consequent danger of overspeeding.

As a mental exercise, the reader may care to consider the extent of any catastrophe when multiple generators are feeding the same busbars, the machines are compounded and have their series fields interconnected via equalising busbars in accordance with the practice of the day adopted for load-sharing. Then repeat the exercise where generator load sharing is achieved by cross or tail chasing, compounding.

Perhaps this is a good time to recount that some problems brought about by the use of regenerative trams were not all catastrophic. Some cases were mildly amusing. The account that follows, and the associated correspondence, appeared in old records of Metropolitan-Vickers at Trafford Park. It can be called

'THE MYSTERY OF THE PHANTOM 'PHONE CALLS'

We must go to Halifax, not long after their prototype second generation tramcar (17) was commissioned and put into regular service.

Complaints were received by the Corporation Tramways Department from telephone subscribers along the route traversed by the car that their telephone bells started to tinkle – not a proper ring – from time to time. The exchange confirmed that they were not trying to put calls through on these occasions, but observant subscribers noticed that it only seemed to happen when tramcar 17 was nearby. The Corporation contacted Metropolitan Vickers about this, who, in turn, got in touch with the engineering wing of the GPO Telephones.

. . . it only seemed to happen when tramcar 17 was nearby.

The area concerned was served by a manual telephone exchange of the CBS1 type (CBS stands for Central Battery Signalling), in which the bell circuit was not connected between the telephone line wires, but between on line wire and earth. The bells had free armatures and were not of the biased type. With conventional trams the rail return current was always in one direction so that voltage drop, and stray current arising, were always polarised one way. Car 17, when regenerating in braking, could reverse the return current flow, associated voltage drop and stray current. Ground potentials which had existed changed in sympathy, and were of such magnitudes that they inverted the standing voltage across the coils of the telephone bells (which were referenced to the earth electrodes at the telephone exchange some distance away). Their hammers therefore were attracted across to the opposite gong to that upon which they were resting, giving rise to a 'ting'. Of course when 17 changed back to motoring the ground potentials reverted to normal and the bells all changed back again. Result: another 'ting'.

The Corporation told Metropolitan-Vickers that it was their intention to close this particular route in the fairly near future, without actually admitting that their track bonding may have been at fault.

Halifax: renowned for its white-knuckle ride hilly terrain but not so well known for its phantom telephone bells.
(Photo: BJ Cross collection)

LOST SKILLS?

Many people forget the importance of the integrity of the return and do not always consider adequately how return feeder should be arranged. That this skill has been lost is well demonstrated today in some (but not all) of the new generation tramways that have been built in Britain during the last decade or so. They rely, for stray current protection, on the insulation of rails from the surrounding paving. However, the effectiveness of these demands levels of insulation which experience shows cannot be achieved. Little wonder that the Board of Trade Rules for the prevention of the 20th century (still in force in Blackpool) laid down three significant voltages: 1½, (one Leclanche cell), 4½, (three Leclanche cells) and 7, which had to be met under all operating conditions. Throughout the nation, in times gone by, they were achieved. The ability of Glasgow to operate successfully a fleet of 40 regenerative tramcars must mean that, behind the scenes, the engineers responsible for the traction supply had made quite sure that line receptivity was never in doubt. Bow collectors on the trams reduced to a minimum the possibility of contact with the overhead line being lost and, where crossings with the trolleybus lines occurred, it was the trolleybus negative, not the tramway positive, that was interrupted.

INTERRUPTED RETURN OF REGEN. CURRENT

Even so, it was realised that a bow collector was still subject to bounce and loss of contact with the overhead trolley wire. The result of this could be an interruption of the return of regenerated current. The effect of this on the tramcar's own electrical equipment would be the immediate initiation of the following very swift chain of events:

a) loss of current through the decompounding series field
b) increase of exciting flux through the armature due to a)
c) increase in generated voltage as a result of b)
d) enhance excitation of the shunt field due to c)
e) repeat of c) and d) until the motor magnetic saturation characteristic prevented the voltage rising further
f) all other electrical equipment on the car becomes subjected to this sudden increase in voltage. This could easily be 20-30% above the norm (or even more) depending upon speed
g) all traction-fed incandescent electric light bulbs that were lit would react by abnormally intense light outputs and be much prone to failure as a result. Any failure could be violent and cause bulbs to explode.

To overcome this problem of occasional loss of connection to the supply network, the regenerative trams were fitted with over-voltage relays that detected the incidence of such an event and introduced a resistance in series with lighting circuits to help protect the lamps from these upward voltage excursions.

This, however, did not address the more general network situation, and if there were only a limited number of other cars available taking the power…….

Consider the following simple scenario. One car is regenerating in parallel and feeding a non-regenerative tram that is hill-climbing, also in parallel. There are no other trams in section and the substation is a rectifier type that cannot accept regenerated power. The motorman on the car taking power throws his controller to OFF. This interrupts the circuit for the regenerating car in exactly the same way as the bow bounce described previously. This time, the rise in voltage created by the regen. car is exported to the other car. Its light bulbs receive the overvoltage but have no protection, and so does the controller which is in the process of opening the traction circuit due to the motorman throwing OFF. The ability of a switching device to interrupt a dc current is approximately inversely proportional to V^3 (voltage cubed). A 20% rise in voltage produces a 73% rise in the duty demanded of the switch. Under these circumstances it is quite probable that the controller will suffer an internal flashover as it fails to make a clean break to the circuit and switch off. Operation of the tram's circuit breakers may also be unsuccessful for the same reason, but then, because of the depression of local voltage the substation will also add to the incident by feeding into the same fault. (The controller flashover had become a fault to earth). Damage to the non-regen. car and its light bulbs was probable, from time to time.

Glasgow's tramway overhead was divided into the usual Board of Trade half-mile sections but, unlike the majority of other tramway systems, often did not put its feeder connections at the section insulators. These sections were centre-fed. The ability to re-allocate regenerated power lay in the feeder and interconnection cables between sections and substations. In Glasgow the regenerative fleet represented only a small proportion of the enormous traction load that Glasgow had. It is possible that Glasgow had some other means of absorbing surplus power such as by means of loading resistors.

A non-receptive overhead line and feeder network was not a hazard-free option with first or second generation regenerative trams though schemes for tackling the same problem created by regenerative trolleybuses of the 1930s were incorporated into the vehicle control equipment. Here, on-board overvoltage detection immediately changed the bus to rheostatic braking from regenerative braking so that the export of excessive voltages was minimised. This neat solution was not readily adaptable to tramways with direct controllers and series-parallel transition.

OPERATING REGENERATIVE TROLLEYBUSES

Those undertakings which replaced their tramways with fleets of regenerative trolleybuses had to address this problem, especially if they also updated their supply equipment to mercury-arc rectifiers. Substations were equipped with resistance load banks that were arranged to be switched in whenever excess power was being returned. Providing that substations were interconnected such as by Hastings Couplers, it was not necessary to install such equipment in every substation. As an example, Manchester had only one load bank, at Ancoats substation, to deal with the whole of the east-side trolleybus routes to Audenshaw, Denton, Haughton Green and Guide Bridge. The load bank was switched in when overvoltage due to excessive regeneration, occurred. Achieving this by a relay and contactor was too slow, however, for satisfactory operation. The rate of rise of voltage could be high. Solid-state electronics was several decades in the future. Apart from the crystal set, electronics was based on valves and vacuum tube technology. Power devices were few, but one – the thyratron – had been developed and was used in this case. It is from this device that the present day thyristor got its name. Its function was similar. A vacuum tube with two main electrodes was open circuit in its quiescent condition. It could be triggered, however, into conduction by means of an auxiliary electrode. Once conducting it remained so until deprived of current by other means. The load bank was thus switched in by the thyratron and then backed by a contactor which closed in parallel, diverting the current away from the thyratron allowing it to return to its normal off state. The contactor would release when the line voltage fell following the completion of the regenerative cycle.

REGENERATION ON CONDUIT TRAMWAYS

Many people are aware that the majority of conduit systems laid down in the late 19th and first quarter of the 20th centuries had both positive and negative 'T' rails within the conduit. This way return currents did not make use of the running rails but used an insulated return conductor.

The polarity of the two rails was determined by the local circumstances, in particular junctions etc., of which the trams were required to tolerate either. Reversal of polarity could take place at the half-mile plough-box section gaps as well as at junctions.

It will be seen that London fitted five cars with trial regenerative equipment around 1933 and no doubt operated them on their conduit system. How could this be possible?

Because the multiplicity of conductor rail gaps was endemic with their conduit, motormen were taught a different technique to those who worked on other systems operating solely with cars fed from overhead lines. In particular they had to 'rush' gaps by acquiring momentum in advance and then coast over them. This had to include branch routes at junctions, making the usual rule of 4

they had to 'rush' gaps by acquiring momentum in advance and then coast over them

161

mph maximum impractical. The important thing to note is the requirement to coast and this would apply to regenerative cars as well.

Having cleared the gap and gained contact with a new set of 'T' rails, the motorman was free to feed up again on the new section. Notch 1 was, of course, resistive, and the current which flowed would be in accordance with the new polarity, the motors having ceased to generate and no longer remained self excited, as a result of coasting. The immediate effect of re-connection was to polarise the motor fields to that of the supply and, in turn, align the polarity of the back emf and generated polarity to match that of the supply.

Should the situation occur where a motorman of a regenerative car did not coast over the gap and a polarity reversal took place, a high current could flow when contact was re-established. This could well trip out all four circuit breakers on the tram. However setting the controller to a resistive notch, eg : first parallel, while the gap was negotiated could reduce the chance of a trip out. Such action may limit the inrush current sufficiently to allow the motors to re-polarise without exceeding the circuit breaker settings.

Even though junctions on trolleybus systems also contained short circuits or polarity change-overs, problems did not arise there with regenerative trolleybuses. They were required to negotiate all junctions at a speed below that at which regeneration could take place in order to minimise the probability of de-wiring.

SOLID STATE CONTROL

The development of heavy current semiconductor devices has made solid- state power control systems a feature of present times and enables static rectifiers to become reversible, a process known as inversion. At first sight it looks attractive to invert power back into the three phase mains but in general, supply companies will have none of this. Their business is the supply of power and not that of a hire shop. Two of the reasons for this are: their network infrastructure has to be rated to carry more power (handling twice, not once) but they would sell less, and secondly they have no control of when, and how much, power is going to be returned. This has safety and stability implications. They will not credit returned power so there is no commercial benefit to the tramway operator. Regeneration's chief advantage lies in displacing power that otherwise has to be paid for.

NEGATIVE BOOSTERS:

Negative boosters subtract from the voltage and were used in earth return systems to keep the potential of all points of the return rail within the Board of Trade regulation limit of 4.2 Volts in order to prevent trouble with electrolysis. Any interest in this nowadays is quite out of fashion!

HE Yerbury, in 1905, gathered from contacting 22 municipal engineers that the average density on feeders up to 1.5 miles in length amounted to 630 Amperes per square inch. Above 1.5 miles this amounted to 527 Amperes per square inch. This occurred under ordinary conditions of service and the percentage of sectional areas of returns to feeders amounted to 81.5% when considering tramway systems supplied with power from a single generating station at 500-550 Volts.

On an average, 3,208 yards of double track were fed from one positive feeder. Abnormal loads, heavy falls of snow or exceptional traffic must be allowed for and the product of Amperes into miles was a simple and convenient calculation of the section of cable required, the length of same and the position on rails of the return conductors.

The average length of positive feeders on 19 large systems was 2,256 yards while on return cables the average length was 1,419 yards. When return cables were of this length a considerable drop of voltage was bound to occur and as the load on return did not fluctuate to the same extent as in feeders it followed that a much heavier loss was incurred than would have taken place if the power station was close to the rails.

Trolley sections of one to three miles were normally fed by separate feeders from the power station and it was advisable to feed trolley-wires at both ends to minimise drop in voltage and thus equalise the potential difference at section insulators. In large systems sections of the line fed by a single feeder were split up as much as possible to facilitate the cutting-out of feeders, or overhead line, in the event of a fault occurring.

With regard to returns, even in the early 1900s, the theories of earth returns vis-à-vis electrolytic troubles had been overrated. The real grievance, if any, should have been directed to nature, or to Mother Earth, and that pitting, or corrosion in lead or cast iron pipes had been caused by the nature of the soil in which the pipes had been embedded.

The Bristol Tramways introduced the first negative booster in 1897 which was designed for a voltage range of 80 -100 Volts. This was most uneconomical and was soon dropped to 40 Volts as being the largest loss considered economical.

Yerbury made example of a 7 mile tramway line taking power from a generator station at its centre as shown on the diagram at the top of the facing page:-

Assuming cars have a 2½ minute headway and run at a speed of 8 mph, from point A to the end of the track on either side there would be 22 cars about 596 yards apart. The total current (at 25 Amps per car) would be 550 Amps. From point B to the end of the line there would be 9 cars taking a total current of 225 Amps. The current from A to B is assumed to be returned by the non-boosted cable near the power station. Four well-bonded 100 lbs/yard rails would have a resistance of about 0.012 ohms per mile and, on this system, there should be approximately a PD(power drop) of 2.5 Volts between the end of track and point B.

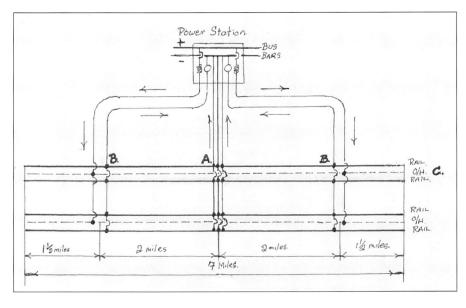

To deal with the current beyond point B two negative boosted cables should be bonded to the rail from half to two-thirds of the distance from the centre of the system to the extreme ends and preferably also two positive feeders should feed the overhead lines near these points so that each booster armature would be excited in proportion to the load. The arrow heads show the direction of flow of current back to the station. Assuming that points B were 2 miles from the power station two cables each of 0.5 square inch cross sectional area would be required to carry the current from the above points (at a density pf 450 Amps/in^2) through the booster armatures to the power station. The voltage required for this load from each machine would be approximately 39 Volts.

The relative length and cross sectional area of non-boosted and boostedfeeders in a system determine the amount of current returned through the booster cables with a given pressure. Where there is a long distance between the power station and the nearest connection to rail there will always be a P.D. between these two points tending to return current to the negative bus bar. This must be borne in mind when capacity of booster, cross sectional area and length of cable for same is determined. When rails are in close proximity to the station, however, with only a small pressure between rails and negative bus bar, the voltage generated by the armature of the booster alone will determine the current returning by the boosted cable.

Here is a simplistic circuit diagram of the principle of the Negative Booster function:

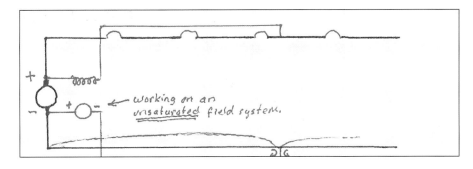

The negative booster only applies to tramway supply. It is there to prevent the rail voltage at locations which were distant to the point of supply from rising as a result of the voltage drop due to return current. It neutralises the rail current by circulating an equal and opposite current through the rail and back to the supply station through the "boosted" negative feeder. It is all about preventing voltage drop in the rails which can give rise to stray currents.

Glasgow's 32 Volt booster would be the maximum voltage that the machine would produce in carrying out this task. It would not produce this voltage continuously, but only as much as necessary to produce the current in sympathy with the current in the positive feeder for the section(s) in question. The actual voltage did not matter as long as the currents were a close match. The booster had to operate as a separately excited machine (from the positive feeder current) over the magnetically unsaturated part of its characteristic.

When regeneration was involved, the nett current returned to the feeder station via the positive trolley wire would reverse excite the booster, reverse the negative boosting current, balancing again.

163

THE ORIGIN OF TRAMWAY REGENERATIVE BRAKING

It has traditionally been the way of tramway historians to give credit for the introduction of regenerative braking on tramcars to J.S. Raworth. Others in the field in early days receive little by way of mention, let alone credit, for what they achieved. John Markham sets the record straight.

Whilst there is no doubt that Raworth was more commercially successful in terms of equipment sold, it would be wrong to make the easy assumption that he was the only technical innovator and the only inventor.

It may come as a surprise to many to realise that Raworth's main competitor used ideas several years before he did, and could be considered the real basis for the second generation regenerative equipments which came along in 1930s Britain. Whilst Raworth was fiddling about with his shunt-wound, series combination only, motors in Devonport, his competitor was racking up the miles on a tramcar with compound-wound motors and series-parallel transitions, in Newcastle-upon-Tyne.

The year was 1902. The competitor was the Johnson-Lundell Electric Traction Company. The Johnson-Lundell Company of Southall, Middlesex, was the UK subsidiary of its American parent. In this the Company followed the same steps taken by Westinghouse with British Westinghouse, and the General Electric with British Thomson-Houston. The Johnson-Lundell Company was established by EHJohnson and Robert Lundell and their early innovations took place in the United States.

It is impressive to note that circuit and motor arrangements used on the Newcastle trials, and results of tests, were still being published in electrical text books at least as late as 1923 on account of the principles they demonstrated, long after references to Raworth had faded.

Johnson-Lundell, in the field of tramcar regeneration, was perhaps the equivalent of Babbage in the world of computation. Others followed some time later (about 28 years or more in the Johnson-Lundell case) with different technology to exploit the principles Johnson-Lundell laid down.

Raworth had the commercial advantage of his association with the British Electric Traction Company (BET), a customer and operator, and with the Brush Electrical Engineering Company, the manufacturer, also associated with BET.

However the Johnson-Lundell Company were innovators in the electrical world – including tramways – for the last decade of the 19th Century, though their products may not now be widely known. Johnson controllers were fitted to the original four cars of the Alexandra Palace Electric Railway in 1898.

Lundell was an electrical machine designer who applied himself to the problem of brushgear sparking, under heavy loads on direct current generators. He invented and patented the split pole-piece with asymmetric circumferentially extended field pole-piece tip to assist in addressing the problem. They featured in many of the generators built by the Company.

In the electric tramway field the Company patented, in 1894, an improvement over previous all-electric surface contact power supply systems. By means of their 'leak circuit' (though in reality it should have been called a 'leakproof circuit') using studs outside the running rails, the possibility of the prime studs remaining live due to surface leakage was eliminated. The system was never laid down commercially in Britain. It will be interesting to see if the divided conductor rail – surface contact system now operatingin Bordeaux would have infringed Johnson-Lundell's 1894 patent.

Evening shadows add to the the pattern of the track and its centre divided conductor rail in the Bordeaux APS system in 2004 – *Alimentation par Soleil* or *supply from the ground.* Opposite is a detail close up of the same.

(Photos. John A Senior)

The Newcastle system was used for service trials with Johnson-Lundell regenerative equipment in 1902.
(Photo: BJ Cross collection)

The Company also dabbled in regenerative battery-electric motor-cars, technical details of which appeared in at least one electrical text book in 1909, and also offered generators for tramway power stations and industrial use.

With his machine background the venture into tramway traction by Lundell is understandable. Whilst he could not use his patent pole-piece tip on a reversible duty motor such as a tramcar requires, he addressed the problem of energy consumption. The introduction of series-parallel transition on two motor tramcars (by others) had halved the energy loss through the resistances, though, due to the number of starts made per day, but the cumulative loss remained significant.

The problem was addressed in two ways:

1. Prime amongst these was the double commutator traction motor, so series-parallel transition was achievable on a single motor equipment. The double commutator motor, though more expensive than a conventional machine, provided a lower cost tramcar (using a single motor) than when two motors were used to obtain the benefit of series-parallel operation. These early double commutator motors had conventional series field windings and were coupled to the axle(s) via sprockets and chains. Axle-hung motors followed.

2. As a further development, field strength control of the motors was obtained using a shunt winding in place of some of the series field.

The effect of combining these two stages was that only one resistance notch for starting was needed, and the compound-wound traction motor was born. It should be noted that its original introduction was on the basis of energy saving during starting, not on the inherent ability such a motor has for regeneration, though the 1902 trials in Newcastle did include recognition of the regeneration feature. The motors included fully laminated field construction to facilitate large and rapid magnetic flux changes, especially those which had to occur at transitions between the three motor combinations. What the Newcastle trial showed, which was not expected, was that the energy loss in the shunt field regulator was almost as much as that lost in the starting resistances on an equivalent series motor tramcar. This, it was realised, happened because the shunt field regulator was dissipating power the whole time the tram was under power, or regenerating, whilst the starting resistance was in circuit only for a small fraction of this time. Johnson-Lundell then changed their approach to claim the energy saving accruing from regeneration gave a net benefit.

Following the 1902 Newcastle trials Johnson-Lundell recruited an electrical graduate from Stockholm Technical High School, Gustaf Lang, to assist in the development of regenerative equipment for tramcars. He realised that the inherently flatter characteristics of the compound motor (and flatter still

of the shunt motor) compared to that of the widely used series motor, rendered it more difficult to achieve an acceptable top speed without the use of many more notches on the motorman's controller – especially throughout the field control range – if the use of the series resistance notches were to be avoided. Without this, notches would be harsh and give a disappointing increase in speed per notch.

His means of addressing this problem was intellectually brilliant but its translation into hardware almost certainly brought about its commercial unacceptability. Double commutator motors were retained but with their field systems modified such that they were field controlled series motors during acceleration and motoring but became de-compounded machines for regenerative braking. The changeover from motoring to regeneration was initiated by the motorman 'pressing a button on the controller handle'. In so doing he operated an additional, remote, electrical switching device, the pole-changer, which had to be introduced to achieve the necessary changeover switching, and this under all load conditions. Lang's development became Johnson-Lundell's 1904 system. A direct consequence for the motor design was that no fewer than 16 leads for each (plus the earth return) were required! (Series motors required but four, and simple compound motors six, plus the earth return).

No wonder it was recommended that the pole-changer was located under the car body between the axles. The pole-changer itself would have been almost the size of another controller, but installed with its cylinder axis horizontal, and probably across the car. Its shaft was operated, and held, in the 'braking' position by an electro-magnet. It was returned to the 'motoring' position, after a braking sequence, by a heavy duty clock-spring mechanism.

The published photograph of the Johnson-Lundell controller suggests that the 'push-button' was not a button in the form of a dead man's handle, but the whole handle was depressed and slid down the shaft as a sliding collar, probably against a spring. This movement was detected by a contact assembly just below the controller top plate and gave the command to the pole-changer.

The equipment involved double-wound armatures (more expensive than normal). Then there was the pole-changer, installation with complex wiring, perceived maintenance costs and additional weight. These must all have stacked up against Johnson-Lundell compared with those of the theoretically less efficient Raworth system. It would no doubt mean that the break-even point for cost benefit would take much longer to achieve with Johnson-Lundell. Even Rawtenstall, where the tramways were managed by the Borough Electrical Engineer, failed to order any Johnson-Lundell hardware for its tramcars (despite what has been published elsewhere). Service trials of Johnson-Lundell equipment, other than 1902 at Newcastle, are elusive, but a reference exists in an electrical text-book to equipment being built for Norwich in 1905, and another to a 'push-button brake' on the controllers of two goods wagon trams built in 1918 for the War Department. (This being the date given in *The Tramways of East Anglia* by Anderson, although 1915 is stated in *Great British Tramway Networks* by Bett & Gillham). These ran on the Norwich tramways (and pulled trailer wagons), to serve as a link between Thorpe Station and Mousehold Heath aerodrome during the First World War. This suggests that Johnson-Lundell pole-changing equipment was being used. But 1918 (or 1915) is too late for such equipment to have been new at the time. It suggests the possibility of at least two car sets previously having been fitted to passenger cars – perhaps about 1905-06 – and subsequently transferred to the powered goods wagons. The tramway company may well have offered to do this in order to expedite the completion of the goods wagons, as the 'official' equipment from BT-H would take some time to procure. They 'officially' had pairs of B T-H GE249 motors and B T-H B49 controllers purchased from B T-H at Rugby. Such equipment was 'officially' transferred to two passenger cars after the hostilities were over and the War Department had no further use for the goods cars, the 'push-button brake' being removed in the process. (B49 controller records make no reference to a push-button ever being fitted by B T-H to control an electric brake).

It is certainly possible that Norwich could have provided facilities to Johnson-Lundell for trial or demonstration purposes:

v   They had several out-of-use trailer cars available
v   The traction electricity supply network was interconnected
v   The Norwich Company was outside the BET Group.

It is a pity that no record of any documented service trial of the final (1904) scheme and equipment has survived.

It is perhaps surprising that Johnson-Lundell did not adapt the application of their ideas, but with single-commutator motors, to four-motor equipments. Burnley, Newcastle and Salford had modest fleets of such cars, and Burnley in particular made savage reference to the cost of power consumed by them compared to its two-motor cars. Newcastle 102, preserved at Crich, started life as a four-motor car.

Four-motor equipments (series motors), with double transition direct controllers, appeared many years later on cars of the Belgian Vicinal.

Full technical detail of the 1902 Johnson-Lundell equipment is given in *Electric Motors* by Hobart, published by Pitman, whilst the principal details of their 1904 pole-changing system appear in *The Electric Tramcar Handbook* by Agnew.

It is unfortunate that none of the Johnson-Lundell motors had interpoles. With wide range field control it would be difficult to operate without sparking at the brushes over at least some of the operating range without them.

It can also be expected that whilst the operation of the two windings of one armature in series would give satisfactory load sharing and acceptable commutation on both commutators, this could

If the flat terrain of the Norwich system was hardly challenging for the Johnson-Lundell-equipped tram, flocks of sheep appear to have created more problems. Tram No.10 is seen at Carlham Road.
(Photo: Online Transport Archive)

not be guaranteed when the commutators were connected in parallel, which occurred at the second transition. Every effort was made to facilitate load sharing in parallel. The same coils were used in both top and bottom windings so they would have similar resistances. It gave rise to another complication – the coil throws had to be different to accommodate this. (The bottom layer was 1-7 and the top layer 1-6). Even so, due to the effect of magnetic leakage, it is unlikely that the field flux penetration of the armature would encompass both windings equally. Poor load sharing would result when the two windings were connected in parallel and it is likely that one commutator would suffer noticeably greater sparking than the other.

Estimated performance curves are shown for comparison. These have been drawn on the basis of likely performance of the two equipments (Raworth R6 and the Johnson-Lundell pole-changing scheme) for which the motor characteristics have been guessed from those of contemporary series motors of similar frame size.

Note should be taken of:

1.   Raworth used resistance notches in motoring over speed ranges 0-5 mph and 6-9 mph. Johnson-Lundell did this over 0-1 mph only. Potential advantage J-L in energy saving during starting.
     (But see the possible fallacy of this argument when we study Maley's 'resistance-less' scheme in the Appendix).

2.   Maximum achievable speed on the level:
     Raworth: 14½ - 15 mph
     Johnson-Lundell: approximately 17-18 mph
     Advantage J-L.
     Birmingham Corporation claimed, in 1906, that Birmingham & Midland Raworth regenerative cars were slow and held up their other services as a consequence.

3.   Use of resistance notches in regenerative braking:
     Raworth: 12 – 6 mph
     Johnson-Lundell: None
     Advantage J-L in terms of potential regen efficiency.

4.   Minimum speed to which regeneration was available:
     Raworth: 5 mph
     Johnson-Lundell: 2¼ mph
     Advantage: J-L.

It seems clear, in spite of the above illustration of the system's potential for greater savings, which the initial cost and perceived technical risk must have been just too much for many potential customers to be persuaded to choose Johnson Lundell.

But Johnson-Lundell principles can be seen being reborn into the second generation of regenerative equipments developed from about 1929 onwards. Whilst Bacqueyrisse in France followed early Raworth with a fleet of 40 or so cars for Paris, in Britain, Fletcher of Metropolitan-Vickers, followed some (but not all) of the features of early (1902) Johnson-Lundell, perhaps without realising this was so. His subsequent technical papers made reference to the early work of Raworth rather than Johnson-Lundell. Later, around 1933, Maley, of Maley & Taunton offered the market his 'resistance-less' scheme of which Birmingham had one (820) and Glasgow had two (698 and 1100) trial cars. This system was a close copy of part of what Johnson-Lundell had been offering as the motoring function of their 1904 regenerative scheme, but adapted for single-commutator motors. Maley's 1936 conversion of Birmingham's 820 from 'resistance-less' control to regenerative braking could be considered direct plagiarism of Johnson-Lundell's 1904 scheme. It at least resulted in an in-service trial of the system that took place for about fifteen months during 1936-37, though under the banner of Maley & Taunton, and for which the then defunct Johnson-Lundell Company does not appear to have received even an acknowledgement. Further details of these developments appear in Chapters 5 and 6 and in Appendix 4.

Johnson-Lundell principles can be seen being reborn into the second generation of regenerative equipments developed from about 1929 onwards

TABLE XXV.—RESULTS GIVEN BY A JOHNSON-LUNDELL DOUBLE EQUIPMENT, WITH TWO 35-HP, 500-VOLT MOTORS, CALCULATED FOR A TEMPERATURE OF 60°, SHOWING THE PERFORMANCE OF THE MOTOR ON ANY CONTROLLER POINT

The table lists, for each Amp Input per Equipment (two Motors) = 20, 30, 40, 50, 70, 100, 160, and for each Controller Position, the following parameters (given here as the table's row labels):

- Controller Position.
- Speed of Equipment in Miles per Hour.
- Draw Bar Pull per Double Motor Equipment.
- Torque in lb for One Motor.
- Torque per Motor in kg at Rim of 887 mm Wheel for 4.93 Gear Ratio.
- Torque per Motor in kg at 1 Meter Radius.
- Speed in Meters per Second at 1 Meter Radius.
- Torque per Motor in Meter-Kilograms per Second.
- Speed in Revolutions per Minute (from Saturation Curve).
- Internal Pressure per Commutator.
- Terminal Pressure per Commutator.
- Total Excitation in Ampere Turns per Pole.
- Series Excitation in Ampere Turns per Pole.
- Shunt Excitation in Ampere Turns per Pole.
- hp Output per Equipment.
- hp Output of One Motor.
- Efficiency in Per Cent.
- Watts Output per Motor.
- Watts Input per Motor.
- Watts Input to Equipment.
- Total of all Losses per Motor.
- Total of Friction Losses per Motor (Gearing, Bearings, Brushes, and Windage).
- Core Loss per Motor.
- Half I^2R Loss (i.e. that of one Motor) in Shunt Circuit, including Rheostat.
- I^2R Loss in Diverter Rheostat for One Motor.
- I^2R Loss in Diverter Spools of One Motor.
- Armature and Brush I^2R Loss per Motor.
- Amp in Diverter Rheostat.
- Amp per Series Field.
- Amp per Commutator.
- Total Amp to Armatures.
- Shunt Amp.
- Amp Input per Equipment (two Motors).
- Controller Position.

Amp Input per Equipment = 20 (Controller Positions 1–8)

Controller Position	1	2	3	4	5	6	7	8
Speed mph	2.8	3.82	4.6	5.72	7.20	10.80	11.66	14.0
Draw Bar Pull	944	900	664	856	316	180	90	—
Torque lb (one motor)	472	450	332	168	158	90	—	—
Torque kg at rim	214	205	151	81	72	41	4.5	6.6
Torque kg at 1 m	18.2	17.4	12.8	6.9	6.1	3.5	—	—
Speed m/s at 1 m	14.7	17.4	24.1	30.8	37.8	55.6	60.8	73.5
Torque m·kg/s	268	303	308	207	231	195	23	41
RPM	140	166	230	286	360	540	580	700
Internal Pressure per Commutator	117	116	115	243	244	492	—	494
Terminal Pressure per Commutator	124	124	124	247	248	494	—	496
Total Excitation AT/pole	6720	2650	2700	4100	2370	7000	—	4300
Series Excitation AT/pole	450	510	730	260	280	730	—	450
Shunt Excitation AT/pole	6270	2090	2090	3850	2090	6270	—	3850
hp Output per Equipment	7.0	8.0	8.1	5.4	6.1	6.60	—	1.1
hp Output of One Motor	3.53	3.98	4.06	2.72	3.04	0.30	—	0.54
Efficiency %	52.4	59.4	60.5	40.5	45.4	38.1	4.5	10.0
Watts Output per Motor	2631	2968	3028	2028	2265	1909	225	401
Watts Input per Motor	5000	5000	5000	5000	5000	5000	5000	5000
Watts Input to Equipment	10 000	10 000	10 000	10 000	10 000	10 000	10 000	10 000
Total of all Losses per Motor	2369	2032	1972	2735	3081	3091	4775	4599
Total Friction Losses per Motor	580	700	950	1200	1540	2400	2910	—
Core Loss per Motor	120	160	260	220	360	900	770	—
Half I^2R Loss (shunt, incl. rheostat)	430	880	480	880	430	480	880	480
I^2R Loss Diverter Rheostat (one motor)	27	36	43	0	9	0	11	—
I^2R Loss Diverter Spools (one motor)	12	16	19	32	39	12	32	—
Armature & Brush I^2R Loss per Motor	200	270	320	67	173	194	37	48
Amp in Diverter Rheostat	9.9	11.5	12.6	5.8	6.3	3.9	—	—
Amp per Series Field	4.4	5.0	5.5	2.5	2.8	2.3	—	—
Amp per Commutator	14.3	16.5	18.1	8.3	8.1	6.1	3.0	4.1
Total Amp to Armatures	14.3	16.5	14.3	18.3	14.3	18.3	14.3	16.5
Shunt Amp	5.7	3.5	5.7	3.5	5.7	6.7	3.5	3.5
Amp Input per Equipment	20	20	20	20	20	20	20	20

Similar columns continue for Amp Input per Equipment = 30, 40, 50, 70, 100 and 160, each with its own set of Controller Positions (1–9). The numeric entries for those groups are given in the corresponding bands of the original table.

RAWARTH TWO-MOTOR EQUIPMENT – ESTIMATED PERFORMANCE

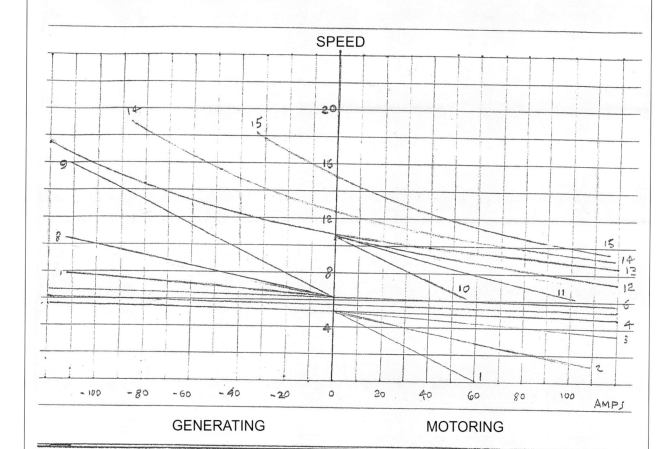

GENERATING   MOTORING

JOHNSON LUNDELL TWO-MOTOR EQUIPMENT – ESTIMATED PERFORMANCE

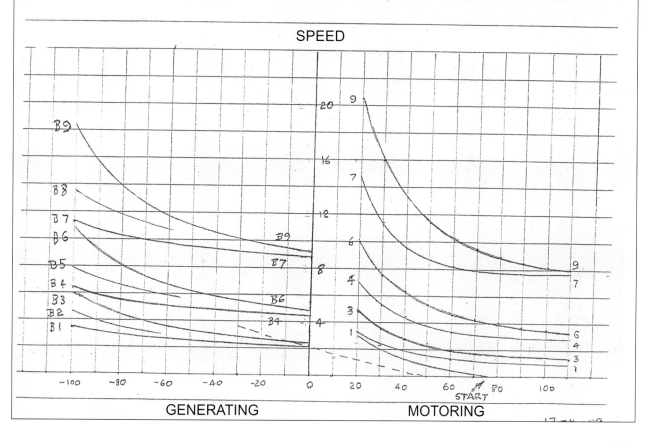

GENERATING   MOTORING

169

THE PROBLEM OF TRANSITION IN REGENERATIVE BRAKING CARS:

Series to parallel transition on conventional tramcars has been set out in Chapter 10, stating the detrimental effects on motors due to overloading during the short-circuiting process. Both Johnson-Lundell and Raworth tackled the problem by opting for open-circuit transition, possibly using the argument that this was no different to the use of a manual gear change on the horseless carriages that were occasionally seen around at that time. In this way was current through both motor armatures ceased during transition but, due to the continuing presence of shunt field excitation, neither armature voltage collapsed.

Johnson-Lundell's laminated field system allowed a rapid change in field flux (increase going from series to parallel, decrease parallel to series). Raworth introduced resistance notches immediately after transition in either direction to act both as a current buffer whilst the field fluxes adjusted themselves, and to provide for ranges of speed which he could not cover purely by field control.

Open-circuit transitions were not seen as acceptable to second generation regeneration's engineers. It probably fell to G.H. Fletcher's Metropolitan-Vickers colleague at Trafford Park, H.K. Ramsden, to devise for the company, a transition scheme which allowed at least some tractive or retarding effort to be maintained, thus minimising the otherwise substantial jerks (motoring or braking) to the tram inherent with the open-circuit system.

The Transition Phase of Regenerative (Compound) Motors:

Ordinary shunt transition cannot be used with regenerative braking control as the short circuiting of No. 2 motor would put a short on an actively generating motor.

A transition resistor must be introduced in this case.

The four stages being:-

(1) Re-insertion of some of the starting resistance
(2) Connection of the negative end of first motor to the return circuit (earth) short-circuiting the second motor in the process, through a Transition Resistor
(3) Transfer of the feed to the second motor from the lower end of the first motor to the top end of the first motor.
(4) Removal of the transition resistor. (This would be notch 6 for an OK45B controller).

Commentary on the Johnson-Lundell 35hp Double-Armature Double-Commutator Tramway Motors:

1. The motors are compound, and used as such in motoring. Both the circuit diagrams and performance curves match this.
2. The description is heavily based upon energy savings arising from field control and double transition – very similar to the later claims of Bacqueyrisse and Maley. Regeneration is almost an aside in this Johnson-Lundell text, and is hardly mentioned.
3. Original thoughts relating to the motor design were clearly wrong. Whilst my point about magnetic penetration of the field flux into the armature is taken, this is addressed not by the use of interlaced windings as I guessed, but by two windings, deep and surface, with different coil throws. I have not come across this idea previously but I do not claim to be a motor specialist. I have reservations as to just how well this would work in practice. It must have been a compromise.
4. From the above winding details given it is clear that my idea of having only one turn per coil would have given major problems with commutator design. Mine would have needed 345 bars available in this size of machine.
5. Fig. 304 implies that there is some circumferential adjustment possible for the brushgear. Note the dotted semi-circle to the left of the top mounting nut. Fig. 311 and 312 *do* have displaced skews approximating to half a slot pitch.
6. Similar coil forms for top and bottom windings would bring their resistances to similar values, no doubt in an attempt to equalise   load sharing when the two windings were operating in parallel with each other. I remain to be convinced just how effective this would be due to the flux penetration problem. Perhaps it is significant that there does not seem to be any reference to how good the commutation was (*ie* the extent of sparking). I suspect that it may have varied a fair bit, perhaps from 'noticeable' to '..... awful', with differences between the two comms. – perhaps one just acceptable and, simultaneously, the other like a Catherine wheel. Quality of commutation is important and is a feature that any potential user would want to know about, especially with a novel equipment.
7. It is very clear that the motor has no interpoles and I thus expect a high probability of even worse sparking during running in the weaker fields, particularly at heavy loads.
8. Despite the up-beat claims in the Hobart text, given the 1905 *Tramway & Railway World* article, and Agnew, one has to interpret that the equipment described in Hobart was not repeated.

9. The system of field coil connections, Fig. 310, can only have been for the Newcastle trial. For the system as described in both *Tramway & Railway World* and Agnew there can have been almost no internal connections between coils. All of the shunt coil connections (convertible turns) and four series coil connections would have to be brought out. Internal connections between diametrically opposite series coils are all that would have been possible.

10. I am not clear whether the series coil, Fig.305, forms part of the compound coil, Fig.309, or was for separate simple series versions of the motor as Figs. 306-308 imply.

11. The photographs of the motor that appeared in the *Tramway & Railway World* feature must have been of a series machine or the compound version of the Hobart text. The latter case could have been the motor used in the Newcastle trials. The photo in Rankin Kennedy could have been a series motor with Fig. 305 field coils.

12. A fully laminated magnetic circuit for the field system was unusual at that date. It makes me wonder if it originates from a desire to make the motor suitable for operation on single phase ac (possibly low frequency, 17-25Hz) which a solid field system would not be. Improvement in terms of commutation would probably not be anything like as great as would be desired and I cannot see it making much improvement in load sharing between armature windings working in parallel. It would help in making the motor less susceptible to flashovers when supply interruptions occurred, such as trolley bounce.

13. Care should be taken in the interpretation of the nine notches described and shown in the diagrams. These are all running notches. There would *have* to be an initial starting notch with a series resistance, as per the *Tramway & Railway World* Article, to limit the current drawn when the car was stationary.

'BACK-ENGINEERING' AS AN ASSESSMENT OF ESTIMATED PERFORMANCE:

The process of 'Back-Engineering' by which assessments of estimated performance are capable of being made to within an acceptable degree of accuracy allow of comparison of machines of yester-year with better judgement. Here John Markham has used his professional skill both to illustrate the propensities of Raworth and Johnson-Lundell regenerative motors and to effect a scientific comparison of the two. The point is formally reiterated that the curves are not definitive but they do confirm an acceptable basis for claiming that the Johnson-Lundell machine was the better of the two. It was its high initial outlay that precluded investment despite probable long-term economy in service.

It should be made clear that the curves illustrated earlier are not to be regarded as definitive, but illustrative only, as there is no hard evidence known to the author to use as a firm basis. They are a graphical guess as to how a typical Raworth equipment may have performed and were produced by taking a typical series motor curve of the day (illustrated in the early edition of *Dover's Electric Traction* but not credited to any particular manufacturer) and re-drawing this on the basis of 25% of the field winding space still being the remaining series field with the rest allocated to the shunt winding. A guess was then made of the value of the armature circuit resistance and a curve calculated that would seem reasonable to comply with the 'about 10 mph in notch 13' quoted in the Rawtenstall Report set out in Chapter 3. The others were filled in by pure guesswork but it is thought that they are reasonable. They certainly show the main features which characterised the Raworth two-motor scheme.

The armature resistance chosen was 0.2 ohms. This dictates the slope of the full field series line. A greater resistance would increase the slope, but also increase the armature losses and reduce the motor efficiency. The flatter the characteristic the closer together notches 4, 5 and 6 have to become.

Attention is drawn to the fact that the guessed characteristics for the resistance notches have been drawn in. Particular note should be taken of the speed range covered by the resistance notches compared with that obtained by the running notches. It indicates that the energy saving during starting would not be as great as Raworth would like people to have thought. Note also that the greatest speed range covered in regenerative braking is also on the resistance notches. This is after reverse transition when notching back to notch 6. Although the calculation has not been done to verify how much energy would be dissipated in the resistances during regeneration it is suspected that it could have been between 25% and 40% of the regenerated power.

In going through the circuits to predict performance it has been noted that the *Tramway & Railway World* write-up does not refer to the change in connections undergone by the series field coils from series in motoring to series-parallel in regeneration. This must increase the number of cables coming out of the motors from 14 to 16, in addition to the return/earthing cable to the motor frame.

A comparison of the curves for Raworth and Johnson-Lundell suggests that the latter was better able usefully to return generated power to the line

A comparison of the curves for Raworth and Johnson-Lundell suggests that the latter was better able usefully to return generated power to the line as series resistances were not used over the main operating speed range. However this needs to be treated with care as in some cases there are hidden losses within the various components (for example, the armature windings) in some combinations, much like the Maley & Taunton resistance-less scheme. A similar point with regard to losses is brought out in the *Tramway & Light Railway World* article concerning the Newcastle tests of 1902. Incidentally the 1902 original Johnson-Lundell scheme was said to be all compound – just like the second generation schemes of the 1930s.

Note that the various tractive effort curves for either Raworth or Johnson-Lundell have not been estimated.

It is clear from the Raworth schematic diagram that open circuit transition was used between the series and parallel notches. It is reasonably certain that the two controller transitions in the Johnson-Lundell scheme would have to be open circuit as well. Operation of the field changer would almost certainly require open circuit action to effect all the changes necessary. In this latter case this would be quite appropriate as the overall effect is to invert the direction of applied tractive effort. Zero tractive effort is a legitimate intermediate stage.

Amongst the many new Light Rail systems in France benefitting from regenerative technology is that at Montpellier, opened in Summer 2000, and currently being extended. Many readers will recall seeing the advance publicity and computer images for the Liverpool tram scheme which was killed-off by HM Treasury in 2006 and which had been based on this French town's transport success. Here one of the smart fleet passes through the pedestrianised Place de l'Opera, formerly the main road between Paris and Marseille, in Spring 2004.   *(Photo: John A Senior)*

Facing page: Germany also continues to upgrade its tramway systems with regenerative technology and in Freiburg passengers can now ride in these all-low-floor Combino cars built by Siemens. Eighteen of these 5-section single-ended vehicles are in use, alongside the earlier cars, and their performance is impressive. Number 282 is seen on the turning circle at the Littenweiler terminus in all-over advertising livery in October 2006.   *(Photo: John A Senior)*

Autumn leaves drop onto the Metrolink line in Manchester's city centre as number 2005, one of the later cars purchased for the Eccles line, runs along Aytoun Street at the end of November 2006 prior to entering Piccadilly. Regenerative braking comes into its own on the former railway sections between the city and the outer termini at Altrincham and Bury, and this car will use part of the former as far as Cornbrook where it will leave the main line and run through Salford Quays and then on-street to Eccles. *(Photo: John A Senior)*

When LRTs came back to the streets of the UK, regeneration finally became the norm in what were, although not always spoken of as such, modern tramcars. We have already illustrated Sheffield and Croydon, but Manchester (opposite) was actually the first, being officially opened by HM The Queen in 1992, Sheffield following shortly afterwards. Below is the West Midlands Centro system, with two cars seen in Wolverhampton. It would not be unfair to say that some of these systems have some way to go before they approach the reliability and dependability of the overseas systems shown earlier. It should be understood, however, that regeneration is not, perhaps, the problem when the word is used in the context of this book as electricity flowing through wires. Regeneration of urban areas seems to be an altogether different issue. *(Photo: John A Senior)*

THE ELECTRICS OF THE REGENERATIVE FLEET OF THE CORPORATION OF GLASGOW

By any standards Glasgow Corporation's Transport Department showed more faith in regenerative control than any other tramway system in the country in the 1930s. 40 trams were placed in service.
Struan Robertson explains how they worked.

THE PATHWAY OF THE MAIN CURRENT AS IT TRAVELS THROUGH THE TRAMCAR:

Regenerative and standard tramcars varied little in their main current circuits apart from the additional shunt wiring of the motors and the extra connections to the controllers. From the overhead – or 'line' – the trolley cable (T-cable) traversed the trolley-pole, bow-collector, whatever, and, on gaining entrance to the car from its roof, ran down the inner aspect of one of the top deck central pillars. This would pass through a protective metal or wooden conduit down to the upper deck floor level from which it would run within the void below this floor to the first automatic cut-out switch in the roof of the platform canopy at No. 2 end, well within reach of the motorman's stance.

From there it re-traced its course to the equivalent position at the No. 1 end platform so that the motorman, at either position, or his conductor on the alternative platform, could replace the circuit-breaker handle and restore the circuit when the overload had passed. The most common cause was the motorman accelerating too quickly. There would be a flash of light accompanied with a bang when the circuit-breaker blew and with impressive nonchalance either of the crew would slap the handle back into position and restore the circuit – to the admiration of the laity at such command. This done, the car would proceed with a little more care on the motorman's part.

From the second circuit-breaker the T-cable descended the bulkhead (or casement) wall to the fuse-box. Below this an emergency circuit was tapped-off via the lightning arrestor to earth through the motor casing, axle and wheels. Any grossly excessive surge in line voltage, such as being struck by lightning, was prevented from damaging access to the car's machinery by a choke-coil below the emergency circuit. This diverted the surge through the lightning arrester circuit. The choke was simply a heavy coil, or solenoid, maybe wound around a wooden core that offered so strong a resistance to a sudden excessive overload that it jumped the lightning arrestor on the emergency circuit to earth in the tramlines instead.

From the choke-coil the T-cable entered the 'bus-bar' cables between the two controllers. These were direct connections between the equivalent terminals in each controller, with the exception of the forward and reverse cables that crossed to the alternative terminal at No. 2 controller to give appropriate directional movement to the car from whichever end it was being driven. From this thick bundle of lengthwise cables, wrapped in a thick plastic tube, feed wires descended to each piece of traction equipment. The controllers housed the necessary switches to route the current to the propulsive machinery. From these negative leads joined the common path through the motor casing, axles and wheels to earth in the older cars while the more modern vehicles had instead spring-mounted carbon brushes which played directly on to the axles. By this means resistance was diminished.

The normal line voltage was originally 500 Volts but the tendency was to rise in most places latterly and 575 Volts was typical. This was unable to overcome the lightning arresters and jump to earth through the emergency earth lead, so continued to feed the motors and drive the car.

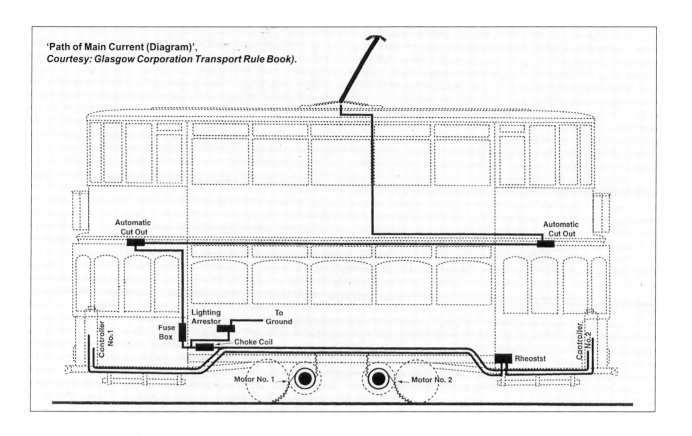

'Path of Main Current (Diagram)',
Courtesy: Glasgow Corporation Transport Rule Book).

Plan of connections of the 'Bus-Bar' cables interconnecting the OK45 B
BT-H Regenerative Controllers on each platform of the tramcar.

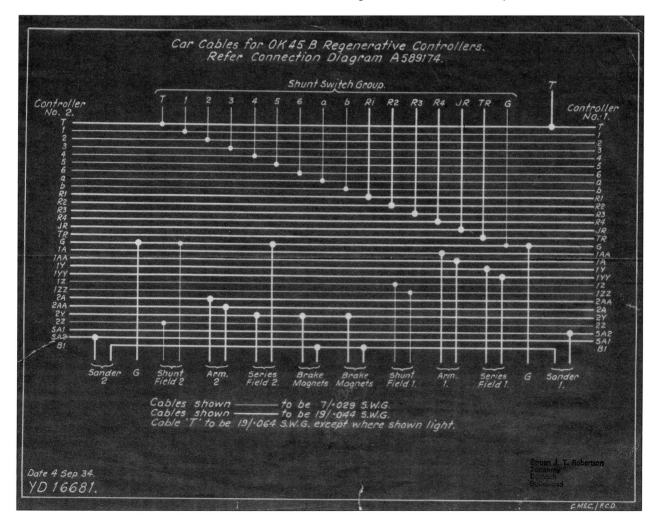

THE CONTACTOR-TYPE TRAMWAY CONTROLLER:

The original tramcar controller, and indeed the controller of most electrical machines, was of the 'drum' variety. The controller handle was attached to a vertical spindle running the full height of the controller down the centre of a drum. On the surface of the drum there were metal contacts in the form of strips and rectangular shapes that were engaged by spring-loaded contact fingers arranged such that when both were in touch, circuits were completed and currents flowed. In other words it was a switching device to route electric currents to the various sites of activity and to withdraw them when no longer required.

As time passed and tramcars grew bigger and faster, carrying greater numbers of passengers, heavier powered motors were required. They consumed heavier currents of electricity to the extent that they were more than the finger-on-drum controller could carry with safety for the motorman. The outcome was the introduction of the contactor-type controller being brought in to cope safely with the much higher power required in service. Metropolitan-Vickers and British Thomson-Houston designated them the 'OK-B' series of which there became ultimately a very big family with many greater uses.

The OK-B type of controller, of roughly the early 1920s, had been introduced to cope with these larger currents and required greater power to break the more powerful arcs which had become dangerous in operation, requiring more insulation. In these controllers, the rupturing of all power circuits had, by careful design, been confined to the small group of eight contactors working in the central compartment on the transverse cam-shaft and they were of robust mechanical and electrical design. They were designed for two or four motor equipments and had four or five series, four parallel and seven braking notches. With further modifications these became the 'Second Phase' regenerative braking controllers of choice.

Forward and reverse operations were obtained simply by throwing the separate reverse handle on to its extreme positions, in either direction. By this means also defective motors could be cut out by the use of extra positions on the same forward and reverse drum without opening up the controller, simply by moving the reverser handle on to intermediate positions clearly indicated for the particular motor. For four-motor control, however, there was an OK-B controller with separate reverse and motor cut-out drums, although it was otherwise similar. Mechanical interlocks between main, reverse and cut-out handles were incorporated to cope with all requirements both for normal running in either direction or for limping back to the depot with either motor– or a pair of motors in a bogie tramcar – out of action.

On the auxiliary drums the fingers were of the hinged type with large contact tips of hard drawn copper to ensure a prolonged life. They could be readily removed for replacement, as were all parts, often by hand, without the need for tools. The finger bars were of rectangular section steel, well insulated, and could be readily removed as a complete unit, or else finger by finger.

The arc-shield was strongly made of fireproof insulation with each contactor being contained in a separate cell-like enclosure. There was a separate blow-out coil for each and there were iron core-plates embedded in each of the barriers to ensure a very powerful blow-out field in the most effective position. Knife-blades on the arc-shield engaged with jaw contacts on the contactor base to energise the blow-out circuits.

Altogether the OK-B controller was a skilfully designed and competent machine

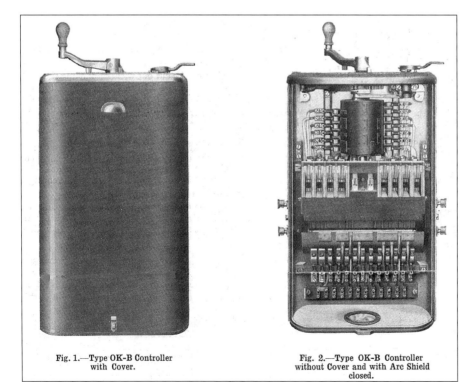

Fig. 1.—Type OK-B Controller with Cover.

Fig. 2.—Type OK-B Controller without Cover and with Arc Shield closed.

The Standard OK-B Controller.

A standard 'OK-B' type controller for a two-motor car, without cover, and arc-shield closed. The forward-and-reverse drum, and spring-loaded finger contacts occupy the top compartment and activated through link-work by the loose, low forward and reverse handle at the right hand side of the top plate. Motor cut-out action was effected through the same loose forward and reverse handle. The main controller handle's spindle passes down through the centre of the forward and reverse drum, to work – through a pair of bevelled gear wheels – the central transverse main power contactor cam-shaft. Below this, in the lower compartment is another transverse finger-and-drum control, for lesser pressure circuits with the main terminal board at the very bottom. This type of controller had to be modified further for regenerative working.
(Photo: The British Thomson-Houston Company, Limited.)

Power and brake connections were made without arcing by a drum that was thrown over by a steel lever on the camshaft into either position as the main controller handle began to move, and remained stationary while the handle was moved over successive power, or brake, notches. The drum was of stout insulation, mounted on strong cast iron brackets clamped to a steel shaft immediately below the arc shield.

The controller casing and cover were strongly constructed of cast iron. The cover was of sheet steel and both were lined with asbestos. The whole controller's compact design was confirmed by its dimensions of 35½in from top to base and its weight of approximately 330 lbs for the two-motor model. Altogether the OK-B controller was a skilfully designed and competent machine, the high points of which could be summarised as:-

1. Negligible wear and tear on the non-arcing segments and controller fingers of the various drums
2. The infrequent and inexpensive renewal of contactor tips
3. Simple and quick routine inspections and rapid adjustments if necessary.
4. Any part, or even a complete controller, could be exchanged smartly without the need to keep a car out of service for any length of time.

Close-up view of Cams and Contactors in one half of the Type OK-B Controller. The mid-line of the machine coincides with the vertical spindle of the horizontal bevelled gear wheel. Arc chutes have been removed.
(Photo: The British Thomson-Houston Company, Limited.)

THE REGENERATIVE BRAKING CONTROLLER — THE TYPE OK45B:

The OK45B controller was produced under two firms' names, of the British Thomson-Houston (BT-H) of Rugby and of the Metropolitan-Vickers (M-V) of Trafford Park, Manchester. Both were taken over in 1929 by the Associated Electrical Industries Limited (AEI) of Trafford Park and each firm was allowed, under AEI, to trade under its own name. They were further allowed to 'pinch' from each other and trade the commodity also under their own names, apparently accounting for the production of both BT-H OK45B and the MV OK45B controllers, and a lot else. Anyway, they were by then both of the same company, AEI, so what did it matter?

The Electro-Mechanical Brake, or EMB Interlock mechanism, was well developed before the Regenerative Braking equipment of the early 1930s. This was incorporated as standard in Glasgow, and was most likely fitted to the controllers at the Coplawhill Car Works. Many other large tramways of the period followed suit. In Glasgow there existed a list of separate dates of the conversion of the 40 Standard cars to 'Regen' control and of the fitting of the EMB Interlock valves to their OK45B controllers. These lists were in contiguous columns on the same table and were, without exception, exactly the same dates in each column throughout. This suggests they were indeed fitted in Glasgow despite the BT-H Company's drawing No. D820580 of the fully rigged out controller suggesting that such fitting had been done in Rugby! So one may have one's choice!

The OK45B was not the first custom-built regenerative controller. There had been several models experimentally equipped for regenerative work leading up to the OK40B type which was installed in Glasgow's Car 305. This was the lone experimental precursor of the production batch of 40 then under construction and discussed previously. The OK45B encompassed all the then up-to-date functions needed for the purpose of compound motor control, in association with the EMB air brake mechanism.

THE MECHANISM OF THE REGENERATIVE CONTROLLER:

From the 'OFF' position, the controller handle was to be steadily and smoothly moved to the third notch. In doing so, the motors would be in series and the resistors were being cut-out, section by

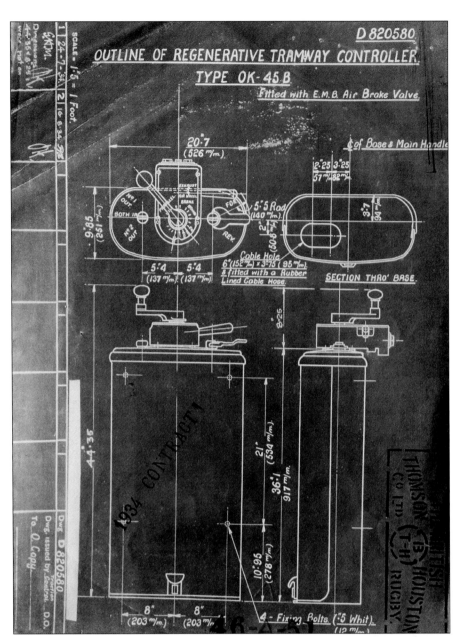

Outline blueprint drawing of the BT-H Type OK45B Regenerative Tramway Controller with EMB Air Brake Valve fitted in superimposition.

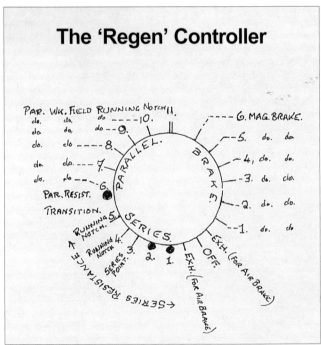

Diagram showing notching and regeneration for regen controller.

180

section from full to zero with the car accelerating to the series point (the 3rd notch) where all resistors were completely withdrawn and the car was running at full magnetic field, equivalent to half speed. This represented the first economical running point, of eight, and the car could be held on this notch in safety, as long as necessary without overheating the resistor banks.

Notes to accompany diagram: opposite

Notching-up:

1-2: Motors in series with resistance in circuit. DO NOT DELAY ON THESE
3: SERIES POINT. FULL FIELD
4-5: Ditto. FIELD WEAKENED PROGRESSIVELY TRANSITION – DO NOT DELAY
6: Motors in parallel, with Resistance in circuit
7: RUNNING POINT. PARALLEL
8: Ditto. FIELD WEAKENED PROGRESSIVELY
9: Ditto.
10: Ditto.
11: Ditto.

Regenerating:

11: Pass back to Point 9
10: Ditto.
9: Regenerating
8: Ditto.
7: Ditto.

TRANSITION – DO NOT DELAY.

Do not pass from Point 7 to point 4 until car speed has dropped to below 15mph

4: Regenerating.
3: Ditto.
2: Pass back to OFF position
1: Ditto.

The next two notches, 4 and 5, were likewise running notches but with progressive weakening of the fields in which the lowering of the applied current diminishes the back-emf of the motors allowing of increments of increased speed of the car. Again, the motorman could rest on either of these notches without damage through overheating of the resistors.

Between notches 4 and 5 there were three steps of transition involving almost full resistance in circuit and where the motorman again had to avoid delay but was required to move his controller handle steadily over this notch where only one small bank of resistance was in circuit. He had to move steadily on to notch 7 where the motors were in parallel without any resistance. This was yet another economical running point upon which he could delay, safely, at leisure.

From notch 7 to notch 11 there was no resistance in circuit and all were safe running notches with the motors in parallel. From the 8th notch to the 11th there was progressive weakening of the magnetic fields so diminishing the internal back-emf and increasing the car's speed. All of these notches used the motors' main field windings only, in clockwise direction.

Thus the eight running notches were 3, 4, 5 and 7-11, inclusive, giving eight, in total, different forward running speeds. The three round notches on the diagram (1, 2 & 6) indicated that it was imperative that the controller handle should not dwell upon them!

When the controller handle was brought back anti-clockwise from 11 down through the parallel notches, the compound element of the motor field windings was brought into action. Regeneration took place from notch 9 downwards, missing out the transition phase, and cutting in again via the series notches 4 and 3 thereby retarding the car's speed. The motorman was required to reduce speed to less than 15 mph when he reached No. 7 notch before he could move smartly to Notch No. 4 when regeneration was re-established with the motors in series, until the next lower notch No. 3 which should have decelerated the car to approximately 3-5 mph After that the application of the air brake (in Glasgow) for a gentle stop, would be used where there was plenty of room on the road. For a quick stop, as might be required in dense traffic, the magnetic, or rheostatic brake would be brought into play to conclude the action rapidly and immobilise the car.

These heavy to handle Type OK-B controllers found little favour with the average motorman who came off duty with aching, or painful arms. However, one section of the car drivers who found pleasure in handling these machines were the members of the Glasgow Corporation Transport team who had joined the boxing section of the Sports Club. They praised the OK-B controllers for developing, and keeping in trim, their powerful left hook!

Do please now turn to the power schematic diagram and derive the pleasure of tracing out the flow of current through the motors in the three main phases of series, transition and parallel in the main current traction circuits, notch by notch. The opening and closing of the main switches from the sequence of switches chart can be visualised, but the 'S', or regenerative switches must be left out until mastering of the main traction circuits is facile. Only then is it possible to enjoy the pleasure of knowing just how the tramcar actually works. It should be remembered that this deals with the Second Phase of regenerative cars of the mid-20th Century. A great deal of advancement has occurred since then.

A warning comes from co-author John Markham who stresses that the 'Dot Charts' or 'Sequence of Switches Charts' (their proper name) are sometimes not entirely accurate and at times misleading. For complete accuracy, power schematic diagrams should be relied upon. Dot charts are infuriating. For instance, they do not make clear how the substitution change between Series Notch 5 and Transition Stage 1 is achieved, in the Manchester Car 420. John Markham would suspect the complete sequence of switches was:-

    (a)   Open R2, then R1

    (b)   Close J, then open JR

    (c)   Close G and M together for Stage 2

    (d)   Open J for Stage T3

The black dots represent the switches in the top diagram that are closed at each position of the controller.

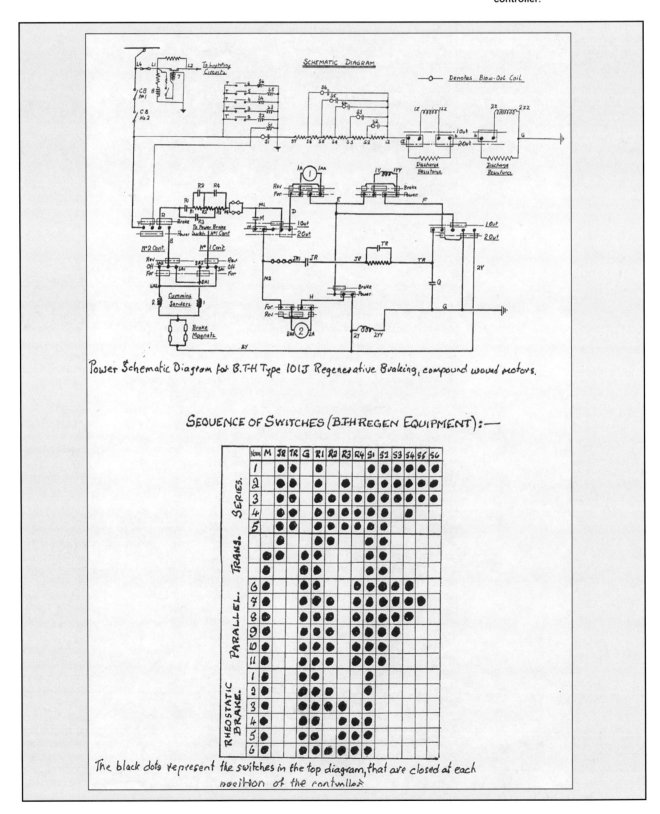

Power Schematic Diagram for B.T-H Type 101J Regenerative Braking, compound wound motors.

SEQUENCE OF SWITCHES (B.T-H REGEN EQUIPMENT):—

The black dots represent the switches in the top diagram, that are closed at each position of the controller.

The BT-H OK-45B Regenerative Braking Contactor Type Controller without EMB Air-Brake Interlock equipment. This has been taken from the BT-H Company Ltd. illustrated spare parts schedule.

The three vertical drums at the top were, left-to-right of the motorman, the motor cut-out drum, the shunt-field drum and the forward/reverse drum. The main handle spindle went down through the shunt-field drum to activate the transverse contactor base, or board, through a pair of bevelled wheels. The square-section bar carried eight steep cams which forced the contacts together against strong return springs which, re-opening, broke the current as the contacts were withdrawn. Below these was the transverse brake-drum, set above the terminal board.

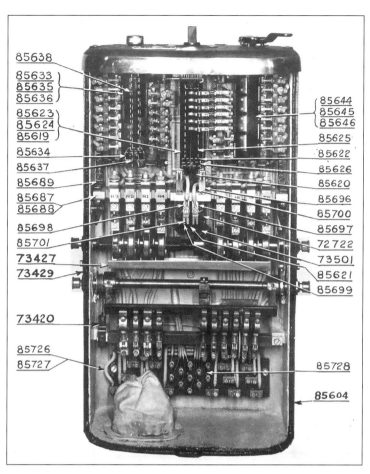

Description Numbers for Illustration of Type OK-45B Regenerative Tramway Controller:

85638	Motor cut-out drum	insulation tube for shaft
85633	Ditto.	M/C drum complete with shaft, segments, spiders, etc.
85635	Ditto.	Insulation drum complete with contact segments only
85636	Ditto.	Insulation drum only (in two halves)
85623	Shunt field drum	Insulation drum complete with contact segments only
85624	Ditto.	Insulation drum only (in two halves)
85619	Ditto.	Shunt field drum complete with ratchet, bevel wheel, etc.
85634	Motor cut-out drum	Spider casting inside drum
85637	Ditto.	Shaft for drum
85689	Contactor elements on board & arc box contacts – C.I. Bracket	
85687	Ditto.	Contactor board complete with contactor elements, switch jaws, supports and support brackets
85688	Ditto.	Contactor board only.
85698	Ditto.	Switch jaw support (right hand)
85701	Ditto.	Special nut for switch jaw bolt
73427	Base details	Insulation tray
73429	Ditto.	Locking plate for bearing nuts
73420	Ditto.	Bracket for brake drum spring attachment
85726	Cable rod for connections – Asbestos braided copper rod 0.252' dia x 2ft 1g (sufficient for one controller)	
85727	Asbestos braided copper rod 0.192' dia x 73ft 1g (sufficient for one controller)	
85644	Reverse drum	Reverse drum complete with shaft, segments, spiders, etc.
85645	Ditto.	Insulation drum complete with contact segments only
85646	Ditto.	Insulation drum only (in two halves)
85625	Shunt field drum -	Insulation tube inside drum
85622	Ditto.	Drum support (fibre) inside drum
85626	Ditto.	Taper pin for drum support
85620	Ditto	Shaft with sleeve riveted
85696	Contactor elements on board & arc contacts – switch jaw support	
85700	Ditto.	Insulating washer
85697	Ditto.	Switch jaw support (right hand)
72722	Main bearing, brackets, shaft for reverse drum & key – main bearing (complete) bolted together	
73501	Shunt field drum -	Bevel wheel
85621	Ditto.	Taper pin with nut & washer for bevel wheel
85699	Contactor elements on board & arc box contacts – switch jaw	
85728	Cable rod for connections – rubber insulated copper cable 0.083' dia x 30ft 1g (sufficient for one controller)	
85604	Base details	CI base

The BT-H Shunt-control Resistor for the OK45B Controller:

This unit is illustrated by the accompanying photographs and circuit diagram. We are dealing with two sets of resistances here, on entirely different circuits: the basic traction resistor of more or less standard design, and the shunt control resistors enclosed in tubes. The units were bulky, divided into a front and back compartment. The latter was for the standard traction resistances and the former for their contactor switches, blow-out coils, relays and so forth. On top there was a pair of banks of special shunt field resistances, each in a special tube. These shunt-field resistances were in two rows of 18 at the top and occupied a similar position at the top.

These type RQ Form G20 resistors were built by the British Thomson-Houston Co, Ltd, Electrical Engineers and Manufacturers, of Rugby, under their reference C100172/1002, and the ohmic resistance values of the traction bank were:-

R1-R2	0.62 Ohms
R2-R3	2.21 Ohms
R3-R4	1.31 Ohms
JR-TR	4.74 Ohms
	8.88 Ohms[1]

In regenerative cars the motor-generated reverse current by-passes the car's resistors and is fed into the overhead to be consumed by contiguous cars in the same section (assuming the overhead is receptive) or otherwise dissipated in the markedly more receptive resistances in the section substation.

In the drawings, and some of the photographs, the following numbers crop up:-

FF is the B.T-H reference to the 'Winding Specification' for each of the various coils marked.
FF21365 are coils for Contactor No. 7 (Type DBR156C)
FF21434 is for the top coil for the Relay }
FF21430 is for the bottom coil for the relay } (Type N. Form H)
Contactors S1-S6 are Type DBR156AM
{ V7500AD Resistance Tube
{ Fuse 'ZED' Type SK1053040 is 4 Amp: 550Volt

The front compartment contained the 7 shunt field operating coils S1-S7 and their blow-out coils, Type N. Form H Relay and Type VAD 7,500 Ohm resistor tube, in the bottom left hand corner, while the top housed the two racks of 18 resistances enclosed in tubes. The Type V. Form AJ shunt field resistances ranged from 22.5 Ohms (S5-S6) to 525 Ohms (a – S2). These may be seen in the respective photographs.[2]

Glasgow's Regenerative Controller Finger-Welding

By 1939 Glasgow had been finding trouble with the OK45B regenerative controllers. These machines were approaching five years of age and had all been very extensively used. The main cause had been the 'JR' fingers which had tended to weld solid, as also had several of the others to a lesser extent (mostly the 'M' and 'G' fingers.)

This would tend to occur because of a considerable rise in temperature, quite quickly, the result of a local hot spot developing when the contacts were closed, and a 'pin-head' arc was maintained across this very small gap. As the temperature of the copper rose it would oxidise and form a relatively high resistance slag at the contact face which in turn would aggravate the heating forming a poor contact that any pin-head arc would naturally try to circumnavigate. This heat, conducted through the whole finger assembly, would soon involve the controlling spring. This would soften, losing its temper, and further reduce contact pressure which reduced its ability to open the contact when the cam-shaft of the controller was rotated into a position where it should release. The spring, which was used for both finger-opening and the controller's knuckling actions, being seated near the contact (the source of the heat) would continue to increase in temperature.

The 'JR' finger being closed throughout series while most of the others were going in and out over that range of activity, would be subjected to a maintained roasting because it would be carrying current for the longest period of time, and so suffer a greatest heating. In the bridge transition it was known that the 'G' contactor would suffer a greater rate of deterioration than all of the others, for the same reason, and the 'JR' finger performed this function in the OK45B controller. The ability of a switch to interrupt a dc circuit is inversely proportional to the cube on the current flowing, so a 20% rise in line voltage meant a 73% rise in the rupturing duty of the controller contacts. It was little wonder that some of them gave up the ghost!

The above description – not taking into account the effect of arcing during transition, could have some other explanation but this one would seem the most obvious possibility. The 'M' and 'G' switches were not far behind the 'JR' switches with regard to the same trouble.[3]

So Glasgow Corporation Transport (GCT) got in touch with the British Thomson-Houston Company and their letter of 1st August, 1939 stated that alterations consisting of merely the rearrangement of one of the existing layout of shunt field resistor tubes would, in all probability, correct the problem. They asked GCT to be good enough to alter one of their 'Regen' car's resistance units to conform to this new arrangement for testing in conjunction with the two motors then being returned from the Sheffield Works of Metropolitan-Vickers. This revision of electrical equipment would then comprise these two motors, together with the shunt field values altered, and the fitting

British Thomson-Houston resistor as fitted to regenerative tramcar.
(Struan Robertson colln)

of 'non-welding tips' to the contactors 'M', 'JR' and 'G'. The alteration of the resistance unit would then consist mainly of the rearrangement of the existing tubes with the addition of two tubes they would shortly be sending forward to Glasgow in preparation for the necessary tests to be performed by them, on site.[4]

The Second World War was to intervene within a month and the results of this test were probably lost in the preparation for more important things to come. JD Markham sums up with:-

> "This seems to relate to problems experienced in service, and shows that BT-H were working closely with Metro-Vick on this project. The reference to 'Our Sheffield Works' relates to the Attercliffe Works of M-V., where GH Fletcher was no doubt overseeing what went on. YD62006 is a BT-H, Rugby drawing, as the shunt field resistances were of BT-H manufacture. However, the controllers were built at Trafford Park, by M-V., and the contact welding problem was one that proved troublesome with their OK range of controllers. The need for non-welding tips on the combination of fingers quoted suggests there were cases of faults, or failures on the road, as a result of welding of one or more of them whilst in service. A problem which has its origins in the fundamental design of the controller was that the contacts were cam-operated to **close**, rather than to **open**. Mr C McCready was a Control Engineer at Rugby"[5]

For the reader-in-depth, a similar letter from Metropolitan-Vickers on the intricacy of the 101J armature coils windings, insulation and banding, together with interpoles treatment, and the MV109 motors for the Glasgow 'Coronation' cars, too long and deeply technical for inclusion here, will have been lodged with the Library of the National Tramway Museum, for care and access.

The Regenerative Motors – the British Thomson-Houston B.T-H 101J:

'In the beginning' – as the saying goes – there were no regenerative motors on the market. They had to be made. The initial experimenters in second phase regeneration had to send their own motors (and controllers also) to the Metropolitan-Vickers firm in Sheffield to be re-wired. As the interest spread in the early 1930s, the M-V firm in the first instance – but followed by one or two other firms – re-designed modern equipment to the regenerative capability as custom-built machines which were naturally streets ahead of the experimental rewiring of traditional gear and of such were Glasgow's BT-H 101J regenerative motors employed in their fleet of 40 regen cars.

The British Thomson-Houston firm had the contract from the Glasgow Corporation (the '1934 Contract') and being by then an integral member, along with the Metropolitan-Vickers firm, of the huge AEI consortium, arranged with M-V to build at least eighty of their MV101DR motors which were by then Glasgow's preferred standard motor, to regenerative configuration and to have their cases cast as 'B.T-H101J' in order to conform with that Contract. The money would all go the same way although 'booked' through the BT-H ledgers! With hindsight, this was the sensible way to have coped with the situation within the AEI kingdom and 'badge engineering' of this type would recur shortly afterwards with the Glasgow Coronation fleets of Mark I and II cars. There were minor structural differences between the 101DR and 101J motors of course, mainly in the armatures, and their bearings as may be deduced from the accompanying tables but examination of the photographs of each show that the only external differences were in the distribution of the entrance ports for the power wiring cables. The 'Regen' motor had these on the non-axle side of the casing while the M-V original design placed them on the axle side of the commutator inspection hatch.

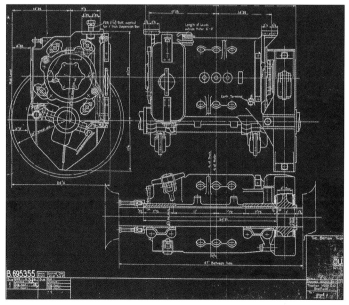

The British Thomson-Houston Co Ltd Rugby outline drawing
of type 101J regenerative braking motor.

A COMPARISON BETWEEN THE MV101 AND THE B T-H 101J TRAMWAY MOTORS:

The MV101 Series

Winding	Wave-wound left hand wound
	Tops to the left 18-19-20-21-22-23
	Bottoms to the right 17-18-19-20-21-22
Number of slots	25
Number of Conductors per slot	6 coil sides per slot (2 in parallel)
Number of turns per coil	2
Step (Embracement of coil)	1-7 slots
Pitch of connections	1-75 (dummy)
Size of wire	0.119' x 0.056' with two wires in parallel (14 such)
Dummy wire (for mechanical balance)	1
Commutator segments	149
Brushes	2
Weight	15cwts 2 qrs

The B.T-H 101J Series

Winding	Wave-wound left hand wound
	Tops to the left 15-16-17-18-19
	Bottoms to the right 15-16-17-18-19
Number of slots	25
Number of Conductors per slot	5 coil sides per slot
Number of turns per coil	3
Step (Embracement of coil)	1-7 slots
Pitch of connections	1-63
Size of wire	0.16' x 0.065 Lewbestos copper wire 5 sep. turns)
Dummy wire (for mechanical balance)	Nil
Commutator segments	135 (some 125)
Brushes	4
Weight	14cwts 1 qr. 20lbs

A Comparison between the B.T-H 101J 'Regen' Motor and the M-V 101DR Example:

This is the British Thomson-Houston Type 101J Regenerative Braking Motor. This photograph was taken after re-winding to series-field control, with removal of the two shunt field leads. Note the current feed ports entering below the commutator hatch on the motor suspension beam side. In the functioning regenerative state there would be six leads entering each motor: 2 traction current leads to the main poles, 2 inter-pole feed leads and 2 main pole exciter shunt wires from the opposite motor's armature not to mention a seventh, to earth.

(Photo: Struan JT Robertson)

This shows the Metropolitan-Vickers Type MV101 motor. The photograph shows the current feed ports entering below the commutator hatch on the axle side of the motor case. Both motors were essentially the same model apart from the positioning of the current ports for, after the 1929 take-over of the two firms by the Associated Electrical Industries Ltd., (AEI), each continued to function more or less independently, contracting in their own names but using each other's products to their own convenience and under their own names.

(Photo: Struan JT Robertson)

The Armature Coils & Connections

The BT-H101J regenerative armature coil was made up of three turns of five rectangular section copper wires, insulated throughout their length, and bound together in three vertical layers. Wound on a former into three turns, the skeleton coil of five wires were bound together, and ready to be further bound, oven-roasted and varnished into the accomplished armature coil, appearing diagrammatically, thus

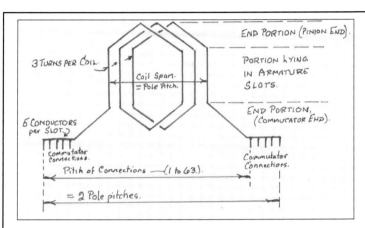

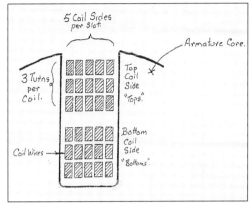

The general outline of the coil being a wave pattern, as seen above, gave rise to the naming of this type of electric motor winding as "wave-wound", or series-winding. This winding provides two complete circuits of the armature and is used for heavy voltage machines, hence their use in transport.

A diagrammatic, cross-section of a single armature slot of the BT-H101J motor, to show the position of the coil wires relative to the slot. Two separate such coil-sides lay in each armature slot, one above the other, their end portions being twisted on the former such that one coil side occupied the bottom of one slot, the twist allowing the opposite coil-side to occupy the top of the slot distant by a coil-step, or embracement, further on. For simplicity, no indication of the insulation of the coils or armature slots has been made.

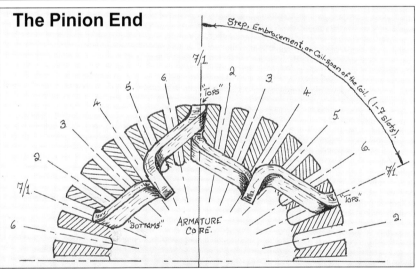

The Pinion End

An end-on diagrammatic sketch, not to scale, depicting the pinion-end of the armature core with two complete coils mounted to show the relationship within the slots to each other, and the diagonal climb to the top position. This equalises the field's magnetic flux effect exactly in the case of each coil, as the armature rotates. The step, or embracement coil, or coil-span, is approximately the same as the pole-pitch of the machine's field. The former-twist of the coils enables its end-windings to cross one another radially without interference.[6]

A diagrammatic sketch, not to scale, of the commutator end of the armature core showing one coil in position and an indication of its onward connections to the appropriate segment or bar.

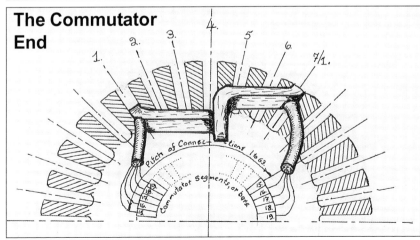

The Commutator End

The Function of the Armature:

The function of this magnificent and orderly bunch of wires is to produce rotational movement out of an electric current. A static electric wire in a magnetic field is inert. Pass a current through that wire and the wire will immediately develop its own magnetic field around itself. As this new magnetic field rotates around the wire, at any moment the wire's magnetic field will be opposing the main magnetic field on one side, as it goes round, and augmenting it on the other, opposite side. This additional magnetic power on one side of the wire pulls, while on the other side of the wire pushes the loose wire, causing it to move out of the main magnetic field in the direction of the reduced magnetic field of the wire, and so to achieve equilibrium. If, instead of a single loose wire, a whole bunch of wires is substituted, made up into an armature, and instead of a single magnetic field there is a circle of magnetic fields made around the armature, the group of wires will, constantly, seek equilibrium outwith the ring of magnetic fields surrounding them and a turning moment is produced in the armature, called the torque, as all the wires continue to be ejected from the various magnetic fields surrounding them in a constant search for equilibrium.

However, the traction motor has four fields around the armature, alternatively positive and negative and this poses the question as to how to reverse the current in the wire at the right moment as it passes into the next field of opposite polarity. This is where the commutator comes into play. The wire has been bent into a number of loops with its two ends brought to attachment to the risers, or lugs, of different commutator segments. As the developing torque rotates around the armature these segments rotate under sets of carbon-tipped brushes which break the current in this wire. As the next commutator segment passes under the brush it feeds it with current that will travel in the reverse direction immediately it passes under the magnet of opposite polarity and the armature receives another kick in the same direction, thus maintaining constant rotation and so rotating the tramcar's wheels. The position of the brushes vis-à-vis the commutator segments is arranged with a view to there being zero current as the segments pass from one or other of the brushes – that is in the neutral position of the fields.

This is a completely rewound MV101DR armature ready for its final revarnishing and re-baking. What had looked like a chaos of loose wires and dangling binding material has now been reconstituted with tremendous patience and skill into a beautiful, smoothly finished machine. This is in all respects generally similar to the regenerative BT-H 101J armature. This particular model had been reconditioned for the Glasgow Corporation Underground trains which, at the time, employed tramway motors.
This photograph illustrating the armature was taken by Struan J.T. Robertson

THE TRUCKS:

There is little to be said about the trucks; in themselves they did not really come into the picture in regenerative braking. The stresses of service, as well as emergency braking, were taken by the motors in regeneration rather than by the brake shoes, and made no difference to the trucks. Their frameworks, carrying both braking systems and the motors, took such blows as they had always done and were designed to do. Certainly, their brake hangings and gear, which only came into the picture during these penultimate few miles per hour of deceleration to rest when in regenerative braking mode, sustained an enormously diminished wear and tear at the expense, now, of the motors and associated equipment.

Glasgow's 40 cars converted to regenerative braking all received brand new standard 8ft 0in wheelbase trucks with 27in diameter wheels conforming basically to the Brill 21E design and they were immediately given the truck number plates off their old 7ft 0in wheelbase semi-high-speed trucks recovered from them in the conversion. These cars retained their new trucks until scrapping although almost all of them had their regenerative motors ultimately converted to series fields. These 'Regen-converted' motors were later used in the modernisation of the Permanent Way Department vehicles that continued through the late 1940s into the early 1950s.

Photographs of both types of 21E-type trucks are illustrated.

This 7ft 0in wheelbase standard truck (No.1039), with a pair of BT-H 101J regenerative braking motors converted to series-field control, was in the early stages of scrapping in the Coplawhill Car Works on 30th June, 1959. Most of the Permanent Way cars on 7ft 0in wheelbase trucks had their Westinghouse 323V 45 hp motors removed and received cast-off 'regen converte''' 101J motors in lieu. Truck No. 1039 came from PWD Sand and Sett Wagon No. 33, seen on trestles in the top left hand corner of the picture, then also undergoing scrapping. These standard 7ft 0in wheelbase trucks were kept for the works fleet of cars.

This 8ft 0in wheelbase standard Glasgow truck is No.1019, complete with MV 101 type motors, was taken from Standard Car 1088 and was awaiting scrapping in the Car Works on 24th October, 1961. The almost identical casings of both the 101J and 101DR machines can be seen. Note, also, the common practice in Glasgow of having different makes of roller bearings to the axles – the SKF variety on the left and Ransome & Marles on the right.
(Photos: Struan JT Robertson.)

REFERENCES: CHAPTER 13

1. The British Thomson-Houston Co., Ltd., Rugby: Drg.No.17956 of 13 July 1934
2. Ibid: Trade Descriptive Pamphlet: No.2673A-6/30
3. Markham, J.D.: personal communication – 22 October 2002
4. Hippisley, E.T.S. – Manager, Traction Sales, per C McCready – The B.T-H Co., Ltd., Rugby: Letter to R.F. Smith, Esq., General Manager, the Corporation of Glasgow Transport Department: 1st August, 1939
5. Ibid – Markham, J.D.: 27 March 2003
6. Arnold, A.: The Modern Electrical Engineer: Vol III : Reprinted July, 1950.: Chap.XIII: 278
7. Ibid: 270.
8. Hirst., A.W..: Electricity & Magnetism: 1947: Glasgow: 2nd Edition: Chap.XV: 352

APPENDIX 1:

THE BOARD OF TRADE ACCIDENT REPORT ON THE HALIFAX COLLISION – 14TH OCTOBER 1904: (A RESUMÉ)

Halifax demi-car 95, driven by John Rhodes, was descending Horton Street on 14th October, 1904 at a slow speed under control of the regenerative brakes, when they suddenly failed. The car ran away out of control for the 102 yards from Cross Street to the point of collision. The gradient was 1 in 14, and from Cross Street for the first 65 yards the track was double, thereafter becoming single and curving round into Church Street at the bottom. The run-away car, 95, collided with car 96, of the same type, that was just about to start from the bottom on the single track. Car 95 was running at an estimated speed of 15 mph and considerable damage was sustained by both cars. 95 continued, out of control, to collide with a dray before stopping.

Motorman Rhodes was severely injured and three passengers in the stationary car were shaken. The driver of the dray was rather badly hurt, being thrown under his cart.

Mr Rogerson, Borough Electrical Engineer, stated the circuit breaker controlling the feeder to the Horton Street section, had come out about 11.45 am, at the time of the collision, and had apparently been a contributory cause of the accident. When Rhodes was descending in Car 95 on his regenerative brake, the blowing of the section circuit breaker would immediately have cut off the regenerative circuit of Car 95, releasing it to run, out of control, down the gradient. Unfortunately the inadequately trained Rhodes was at a loss being ignorant of the use of the rheostatic emergency brake that would have brought the car under control. His car was running at an average of 8 mph, which, over 100 yards, would have taken 26 seconds before the crash, leaving him very little time to think what to do, and act.

Lt Col Druitt, RE, later examined Car 95 and found both the regenerative brake and the emergency brake functioning normally. He chided the Halifax Corporation for the absence of run-back brakes, rendering the Corporation non-compliant with the BOT. Regulations for the Halifax Corporation Tramways No.1(c) in this respect.

John Markham offers the following commentary on Colonel Druitt's Report, postulating a rather different conclusion:

Lt Col Druitt . . . chided the Halifax Corporation for the absence of run-back brakes

> It is well known in tramway circles that the tramway system of Halifax had more hills and undulations per route mile than any other in the Kingdom. It was also one of the first to see the possible benefit the use of small, one-man, demi-cars could have on routes with lower traffic potential. As seen in Chapter 2, their two demi-cars incorporated two traction motors unlike Southport's. Halifax's gradients made this a necessity. Whilst there is nothing unusual in general about using two motors, it gave Raworth the problem of ensuring reasonable load sharing between the two motors while regenerating. Shunt wound motors have a flat characteristic and as such do not have the inherent ability to share load in the manner that series wound motors do; as seen in Chapter 5, this was re-exploited in Paris in the late 1920s.

During car 95's descent of Horton Street in the charge of Motorman John Rhodes, it was being held in check by the regenerative brake. When the automatic circuit-breaker on the power station switchboard for that section of route tripped out the power supply was disconnected and there was no longer anywhere for the regenerated power to go. The retardation caused by regeneration immediately ceased and the car began to accelerate down the incline.

Motorman Rhodes repeated the routine for establishing regenerative braking, unfortunately to no avail and the car continued to accelerate. He then applied his handbrake but was unable to avoid colliding with the other demi-car 96 which was starting to come the other way on the single track line which Rhode's car had run on to.

The controllers did not have rheostatic braking notches but did have an 'emergency brake' position on the reverse handle. The enquiry established that Rhodes was unaware that the controllers on the car had this feature which, had it been operated, would have provided a powerful short-circuit brake on the car. This was later proved to be in working order. In addition the car had a hand operated slipper brake which Rhodes did not use.

Colonel Druitt drew attention to the failure on the part of the management to ensure that their motormen were properly instructed, and also that there was in place a Board of Trade Regulation in force in respect of Halifax that their trams had to be fitted with a run-back brake, which the two demi-cars were not. It seems likely that Halifax also realised that this was true for others in the fleet, and implemented a programme of modifications to other tramcars in their fleet to achieve this (though this is not part of the regen. story).

The report into the Horton Street runaway of car 95 also contains details of the controllers fitted to the car at the time. The features described do not match those of the R6 controllers made for Raworth by British Westinghouse, (their type T1R). They are much more akin to the H.G-2 type of controller made by Brush, which was, in turn, a derivative of their H-2 type.

the power supply was disconnected and there was no longer anywhere for the regenerated power to go

Detailed technical information on Brush tramcar traction equipment is scant compared to that of other manufacturers and, unlike other suppliers, no Brush traction hardware is known by the authors to have survived the 1930s, never mind reached any stage of preservation.

Nevertheless, using a known description and circuit of the H-2 controller, and the details given by Colonel Druitt, it is possible to suggest likely forms of construction, working and operation of the H.G-2. Confirmation (or correction) of the details postulated below would be welcome.

The H-2 was a simple two-motor series-parallel controller without graduated rheostatic braking, but with an 'Emergency brake' facility obtained by putting the main handle to OFF and pulling the reverse handle from FOR through OFF and REV into an emergency position. In this respect it was similar to BT-H 'K' types and EEM 'DE1' types. However, unlike its competitors, the H-2 did not have an equivalent fifth position for the reverse handle giving reverse emergency braking.

It is stated, in a general description published in 1911, that the H.G-2, when in motoring notches, used the same connections as the H-2. Regeneration was obtained by moving the reverse handle 'To change the motor connections from motoring to generating'.

This sets us some problems; in motoring and emergency brake the motors would be pure series wound machines, but the change to regeneration would require them to become shunt wound machines, with their armatures connected permanently in series, or become compound wound which in turn would have required a different form of series-parallel transition to the H-2 if parallel were used.

As regenerative braking on car 95 is described by Druitt as an effective holding brake on long downhill gradients the shunt wound option using the series notches only seems much the more likely. If this is so, each motor field would probably need to have two coils per pole (c/f Johnson-Lundell), with the coils in parallel (motoring) and series (generating) per motor. The controller adaptation from H-2 to H.G-2 would then be confined to the reverser barrel and its finger-bar, with the reverse handle gaining a fifth position; the clockwise sequence becoming:

> Forward regeneration
> Forward motoring
> OFF (where the handle may be removed)
> Reverse motoring
> Emergency brake (forward motion only)

The usual interlocking would be provided requiring the power handle to be in the OFF position before the reverse handle could be changed in its position, but, in addition, there would be an additional restraint on the power handle preventing progression beyond the full series notch when in the 'Forward regenerating' position; much in the same way parallel notches are denied on many controllers when working in the motor cut-out position.

Now comes the tricky bit – what the motorman would have to do to use the regenerative brake.

The procedure with equipment as described above is easiest understood if we assume the car to be stopped at the top, but on the downhill slope. Before starting the car the motorman would move the reverse handle anti-clockwise one position from FOR to REGEN; this would prepare the motor fields for being excited from the line as shunt wound machines. Leaving the power handle in OFF he would allow the car to start rolling under gravity and then feed up through the usual series resistance notches until by full series he had the car running downhill at about 6 mph the Board of Trade speed limit on the Horton Street gradient, and regenerating into the line. Notching back would introduce resistance into the motor circuit and allow the car to run faster, so, in order to stop the car, he would have to apply one of the friction brakes to take over as he notched back or threw OFF. (Note: the horizontal axis is motor current in amps: vertical axis is speed, mph, and the intersection by the regen. curves of the speed axis will occur at around 4½ mph in the case in point).

It would be possible to perform the change from motoring to regeneration and vice-versa on the move, providing the power handle was in the OFF position. Using regen. for service stops would be cumbersome and not particularly efficient where speeds over about 8 mph were involved, and it would only be effective down to about 5½ mph

It would be possible for a motorman to start the car from rest in the REGEN position, but its response would be expected to be jerky, the speed not exceed about 4 mph on the flat, and parallel notches would be denied him.

It seems possible the H.G-2 controller may have been developed as part of Raworth's work at Devonport. The lower set of curves at the end of this Appendix show the likely form of series notch characteristics had the motors changed from series wound (motoring) to compound wound (regen.) Note in particular the regenerated current would have been less due to the demagnetising effect of the series winding, and as such probably would not have produced sufficient retarding effort to act as a 6 mph holding brake down the 1 in 14 gradient of Horton Street. In both cases (shunt wound or compound wound motors), it should be noted that "notching up" in the 'REGEN' position through the series notches produces an increase in tractive effort, motoring or an increase in retardation, braking, depending on the car's speed. Confusing?

The above suggestion of how the H.G-2 controller was used prompts the thought that confusion at times of stress could occur, and in part explains why, when Motorman Rhodes lost retardation from regen. braking, it took a significant time for him (using both his hands – one on the power handle, the other on the reverse handle as one would assume) to repeat his attempt to apply it; consuming much of the 26 seconds available to him (as calculated by Druitt) before the first collision.

It is possible for us to consider another scenario which takes in another feature of the incident about which Colonel Druitt makes observation but no comment, though he may well have explored it.

If Motorman Rhodes had forgotten to change the reverse handle from FOR to REGEN when he started down the hill, his notching up to engage the regen. brake would instead act in motoring mode and accelerate the car. Upon realising this his likely reaction would be to throw OFF, then move the reverse handle into the REGEN position and apply the regen. brake as quickly as possible to recover the situation. In a state of anxiety he probably did so by bringing the power handle to the 'Full Regen.' position ('Full Series' on the controller), without pausing on the intermediate resistance notches. The chances are that the car's speed would be well in excess of the usual 6 mph At 10 mph the generated voltage of the two motors in series would be in the order of 840 Volts, and on striking the full series notch this would become paralleled to the power station busbars at only 500 Volts. For one fleeting moment regenerating car 95 would be powering all the tramcars in Halifax! Under these circumstances it is no surprise that the feeder circuit-breaker for the Horton Street section tripped out on the resulting momentary current overload. Whilst this is speculation, no-one else seems to have offered any explanation for the tripping out of the feeder circuit-breaker at exactly this crucial time.

For one fleeting moment regenerating car 95 would be powering all the tramcars in Halifax!

The question may well arise in the minds of readers 'Why didn't the circuit-breakers on the tram trip as well?'. The probable answer is that there weren't any. Surprised? It should be remembered that demi-cars were intended to be cheap to build and operate. Provision of two non-automatic canopy switches and a traction fuse was a cheaper option than circuit-breakers.

Traction fuses of the day were crude in the extreme, and often comprised a replaceable single wire element of adequate diameter trapped under two wing nuts in a box lined with a refractory material (usually asbestos based). Such fuses would have a relatively long pre-arcing time whilst the element (often tinned copper) heated up to its melting point. A circuit-breaker with a magnetic strip system did not suffer this thermal delay inherent with a fuse. Under the circumstances of the Horton Street incident it is quite reasonable for the power station circuit-breaker to respond more quickly than a tramcar fuse to a sudden, but short, overcurrent. A fuse, hot but not melted, would be intact.

Postscript:

Following this runaway, and the matter of the lack of run-back brake (even though this had no bearing on the accident) Halifax removed the Brush H.G-2 controllers from cars 95 and 96 and replaced them with pairs of BT-H controllers type K.10.

By so doing 'reverse emergency brake' became available on the reverse handle by pushing it beyond the 'Forward' position instead of 'Regenerative brake'. The result would probably serve as a run-back brake to the satisfaction of the Board of Trade Inspectors even though graduated rheostatic brake notches would not be introduced, nor would the cars be able to have magnetic track brakes.

Cars 95 and 96 would thus cease to be regenerative, and if the original Brush 'Raworth' motors were retained they would have to be permanently wired as series motors.

The decision by Halifax to reintroduce regenerative braking only a few years before abandoning its trams completely speaks volumes for the savings that a hilly system could make by reducing the amount of power consumed by its rolling stock. *(STA)*

APPENDIX – 2

THE OFFICIAL REGENERATIVE BRAKING DRIVING INSTRUCTIONS

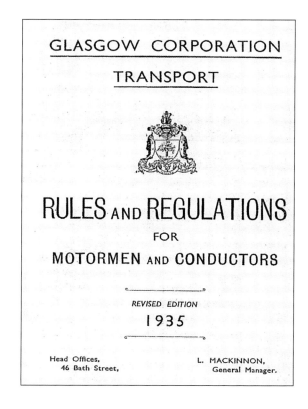

GLASGOW CORPORATION
TRANSPORT

RULES AND REGULATIONS
FOR
MOTORMEN AND CONDUCTORS

REVISED EDITION
1935

Head Offices,
46 Bath Street,

L. MACKINNON,
General Manager.

REGENERATIVE CONTROL CARS

The General Manager finds it has become necessary to remind some motormen that it is absolutely essential to operate regenerative equipment according to instructions.

Especially have cases been noted of motormen using the resistance notches for running, and coasting on the off position.

To run on resistance notches or coast on the off position and then effect braking by working up on notches 3, 4, and 5 is entirely contrary to instructions, and not only defeats the purposes of the equipment, but inevitably leads to its serious damage.

Rigid adherence to instructions as laid down in the Rule Book is necessary if the comfort, safety and economy which this equipment make possible are to be achieved.

To facilitate maintenance, motormen are particularly requested in any case of breakdown to note in report:-

(1)   Whether the car was being driven from No. 1 end or No. 2 end.
(2)   In what direction the car was travelling – to which terminus.
(3)   How the car behaved leading up to the trouble.

If repeated blowing of automatic circuit breakers occur, especially when starting, each motor should be cut out in turn, if necessary, and the car run to the depot on the good motor. Not too frequent blowing of automatic breakers should be allowed before resorting to cutting out motors.

L. MACKINNON,

General Manager

46 Bath Street,
23rd July, 1935

(TO BE RETAINED IN RULE BOOK)

REGENERATIVE CONTROL

The operation of regenerative control equipment is entirely different in some respects from the operation of ordinary equipments. Whereas with ordinary equipments the controller handle must always be moved smartly with one movement to the "off" position from any power notch, it must, generally speaking, only be moved back one notch at a time with regenerative control.

The controller has eleven power and six magnetic brake notches, the power notches comprising five series and six parallel notches. Only the first and second series and the first parallel are resistance notches; all others, excepting top parallel, for a reason which will be explained later, are to be considered economical running notches, inasmuch that when working back or coasting on these notches the motors are generating and feeding current to the overhead wire. When starting up with a clear road the controller handle will be taken as usual over each of the five series notches, then quickly over to first parallel, and rather more quickly than usual over to 6, 7 and 8, and again as usual over 9 and 10 to 11. The 11th or last power notch should not be used for more than a second or two, as little regenerative effect is produced there and, in any case, the speed is greater than is often required.

Having reached the top notch the motorman may find it necessary to reduce speed; this is accomplished by moving the handle back to the tenth notch or further, one notch at a time. This will give a gradual reduction of speed if the handle is moved properly.

At each step, excepting on the three resistance notches – 1st parallel and 2nd and 1st series – regeneration is taking place and current is being fed to the trolley wire.

The handle should only be moved back over the notches far enough to effect the required reduction of speed, and may immediately be moved up over the notches again from any position if increased speed is required.

While with ordinary equipments full advantage must be taken of a car's ability to run for long distances on the 'off' position, so effecting, in a negative way, economy of current, with regenerative control the advantage lies in the regenerative effect obtained, while braking, by working the handle backwards on the power notches. When making a stop, the handle will be worked back notch by notch as directed to the 'off' position. The speed will have been so much reduced by the regenerative effect on the armatures and wheels that a light application of air brake will bring the car to rest.

Just the necessary amount of air pressure will be used to hold the car pending the re-start, and the usual care must be exercised to ensure a perfectly smooth start.

The procedure outlined of working backwards over the notches must be rigidly adhered to at all times unless an occasion arises when the motor-man decides that the very effective braking so obtained will not suffice for a particular emergency.

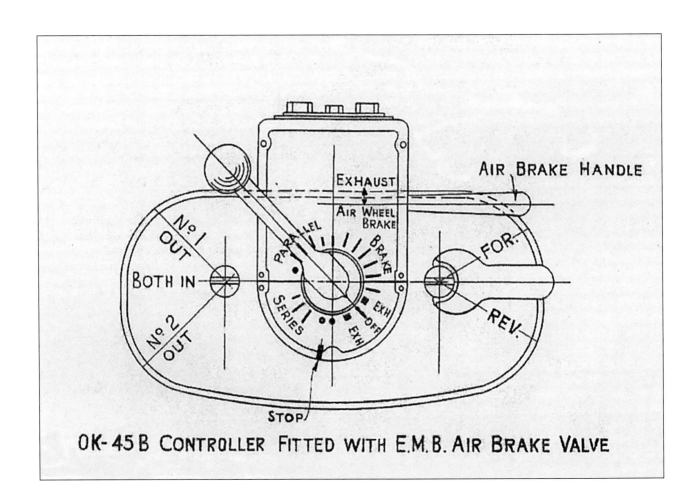

OK-45B CONTROLLER FITTED WITH E.M.B. AIR BRAKE VALVE

In such an event the controller handle may be switched smartly to the 'off' position, and then on to the magnetic brake notches for the emergency stop. Occasions for the use of the magnetic brake on a car with regenerative control should be extremely rare.

If for some reason it has been considered necessary to switch quickly to the 'off' position, and it is desired to 'feed up' again without reduction of speed, the controller handle should be brought quickly to the power notch corresponding with the speed of the car.

SECTION INSULATORS

To prevent arcing at section insulators, it is necessary to produce the condition, that power is neither being taken from, nor fed into the trolley wire in passing. The procedure to be followed depends on the notch in use when approaching the insulator.

If the handle is on a regenerative notch, it should be moved back one notch, and then up one notch. The two movements – back and up – should be made just before the bow collector reaches the insulator.

When approaching an insulator, the motor-man should endeavour to have the handle no higher than the ninth notch. It is imperative that the handle should not be moved back to a resistance notch and then up again.

If one of the first three notches is in use, the handle should be brought to the 'off' position.

DISCONNECTING OF MOTORS

When a motor is disconnected, the regenerative action, and the magnetic brake, are still effective, although not quite as powerful as with two motors. The cutting out of a motor is accomplished with the aid of the 'reverse' handle.

The controller top has two projecting spindles, the one on the right is the normal position of the 'reverse' handle for the direction of the car. The one on the left is for the purpose of cutting out motors by the use of the 'reverse' handle. In the forward position No. 1 motor is 'cut out,' and in the backward position No. 2 motor is 'cut out.' When a motor is 'cut out' the controller handle will not pass the top series notch. The need for disconnecting having arisen, the motor should always be disconnected at both controllers.

Newcastle tramways were very early in the game of regeneration, but unfortunately there is no record to confirm which car was involved. It might just have been one of these!
(SITA collection)

APPENDIX 3

AUXILIARY EXPERIMENTATION
FIELD DIVERTER, WEAK FIELD CONTROL
ETC.

Ancillary experimentation related to regenerative braking in Glasgow must be mentioned here as of flanking, but direct, interest in order to keep up with the overall picture of that time. This resolves itself into an almost bare mention of known fact, for very little information is reclaimable at the present time.

Weak-Field Control:

Weak-field control, or Weak-Field Diversion, was offered by the American General Electric Company on some of its earliest designs of "K"-class controllers in 1898, but was quickly abandoned as the motors with weakened fields were totally unsatisfactory due to magnetic scatter. It was not until interpoles became available somewhere around 1911 that this problem came under control. It was especially, and particularly on heavy loads, involving poor, to very poor, commutation in these early machines that such trouble reached its worst.[1]

At constant voltage the speed of an electric motor is in inverse function of this field strength: reduce the field strength, therefore, and the tramcar will go faster! That is the principle of Weak Field Control. Incidentally, the tractive effort is lowered by the process of field-weakening because the torque of a motor is a function of the product of the armature current and the field strength.[2]

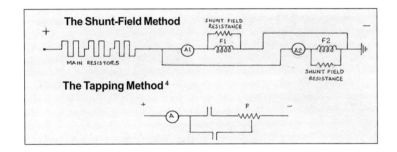

The Liverpool streamliners suffered an enforced reduction in top speed capability after manager Walter Marks had the weak field facility removed. *(RJS Wiseman)*

At full speed, in both series and parallel top notches, it was long known that by diminishing the motor's field strength, an extra spurt of speed could be obtained to the extent of around 15 and 30% in two increments, by eliminating some of the field turns through the introduction of a shunt field resistance, or by the use of field tappings.[3]

The uppermost diagram (facing page) illustrates Weak-Field control in top parallel, with shunt field resistors across both motor fields, and the main traction resistors cut out. These shunt field resistors 'cream off' a portion of the current through the motors' fields thus reducing the motor-fields strength. As the motor-fields are weakened, the motor's Back-emf drops, allowing an increase of the contra-flow of applied voltage and therefore an increase in speed. Literally the weakening of the motor's field strength increases the speed of the motors. The lower diagram indicates the use of contactors to cut out portions of the motor's fields in order to weaken them.

If the motor field windings have single tappings then four economical running speeds become available. There are two speeds in both top series and top parallel, corresponding to full field and tapped field in both series and parallel. Should two tappings be provided, six economical speeds become available. All these weak-field tappings are economical running notches.

A tapped field motor would require its field coils to be wound in two sections and an extra lead would be required to be brought out.

"My guess is that the OK20B controller required external contactors to provide the 'divert' (or Tap-change). There was little chance that the cam-operated contacts within the controller could have been used for field-weakening. Eight was the maximum that could be accommodated within the controller casing and these would all have been required for notching and transition forbye any tap changes. The only information I have on this type of controller does not indicate the number of notches provided. However, they were supplied without main power handles, implying that air-brake interlock valve boxes were fitted. Unfortunately I do not have a diagram for the OK20B controller".[5]

The transition from full field to tapped field was effected by short-circuiting a portion of the motor's windings and cutting it out of circuit, but this process could subject the motors to large rushes of current, detrimental to the machinery. Liverpool found this to such an extent that it removed all the weak-field apparatus from its cars. However, this rush of current was ultimately controllable by inserting an intermediate step between full field and tapped field consisting of a resistor in parallel with the designated section of the motor field winding for removal, followed by the cutting-out of both the designated field section and the resistor altogether. This was found to give an altogether much smoother transition event.

Blackpool's streamliners, from a slightly earlier period, also had the weak field removed, but, unlike the Liverpool cars, lived to see another day – in fact many! *(IGS collection)*

Motors were all low resistance equipment and for starting, the supply voltage had to be brought down (by the traction resistors) to such a low value of voltage.[6] Once going, if a running series motor, part of a series circuit, is short-circuited across armature and field, as in transition or in rheostatic braking, it first of all generates a flash of heavy reverse current, opposing the generated

voltage and the machine is rapidly demagnetised, becoming electrically inert. The magnitude of the reversed current is considerable, in Amperage, but its voltage is very low to the amount of energy is limited and the duration of this transient condition is extremely short – generally less than 0.4 seconds! So, the change-over from series to parallel, due to this characteristic of the series type of traction motor, termed the 'Short-Circuit Transient', has to be so quick as to be almost instantaneous.[7] (The theory of the Short Circuit Transient is also discussed in Chapter 10).

A renaissance in Weak-Field Control interest was to occur during the period of the modernisation of the early 1930s, in Glasgow, when some twenty or so recorded cars including both Standard cars and the so-called 'Kilmarnock Bogies' were field-divert fitted.[8] Where MV OK26B controllers were recorded, these might represent cars so fitted after removal of the OK20B variety.

These diverters seem to have been in service for quite a long time judging from the very few official dates of their removal. London had very many – cars 785-788 of the E/1 class, among many other London company cars, had weak-field equipment including the entire 'Feltham' fleet, and many of their predecessors in both the MET and LUT fleets. Birmingham Corporation had one or two sets and very probably many other systems may have experimented with them. It would be very difficult to attempt to produce a definitive list, for many of the country's tramway concerns were petering out due to motor 'bus competition and such details were very rarely recorded. Blackpool's 'Balloon' double-deckers, and their single-deck cars of the 'Railcoach' type, all had manual weak-field, in parallel, when new although most (if not all) had it removed fairly early on in life. Liverpool's bogie streamliners were fitted with field diverter equipment when new, although again, this was manually controlled by the motorman and led to a problem with overheated interpoles on the traction motors as referred to previously. It was removed from most of the fleet although there are stories of odd cars retaining it or having it reinstated surreptitiously!

The difference between 'Tapped Field' and 'Weak Fields' has been gone into, but simply it lies in the forms that motors required their fields to be wound, in two sections, and an extra lead would be required to be brought out. The Glasgow controllers that were fitted with weak-field diverts were of the OK20B type predominantly, although perhaps there were two or three of the OK26B type (if it were not that they were changed to the OK26B type when 'de-diverted'). The OK20B type was, most likely, a predecessor in design to the OK26B but lacked the latter's feature of a 'quick-break' assembly in the 'R1' and 'S' positions. This was a feature that Metropolitan-Vickers claimed as a 'Lineswitch', otherwise they might well have been electrically the same. If so, the 'weak-field' arrangements cannot have been a notch on the controller but would have had to be automatic, in top parallel notch – something akin to the Igranic Auto-shunt referred to by Lawson in his Birmingham Rolling Stock Book.

There were several advantages in the use of weak-field control, outstandingly in congested street work for getting rapidly out of a traffic snarl as it eased up; also in peripheral sections of scanty trade which demanded two very different working speeds. Low city-centre speeds in crowded traffic and closely spaced car stops could be served economically by operating the car's motors under full field while the faster suburban stretches, with widely spaced stops that allowed longer spells of fast running, could be maintained by running on the weak-field notches. This achieved economy of energy consumption while securing faster speeds.[9]

The introduction of weak-field control in existing cars, however, was unlikely to lead to their economy in power consumption. The car performance would be enhanced by an increase in top speed and the car would be accelerated to this speed at a greater rate than by allowing it to accelerate on the top parallel notch, thus more power would be likely to be consumed.

The real economy benefit of field control came only if the overall car performance remained the same and field control was used to displace some of the resistance notches – claims made by Raworth, Johnson-Lundell and Fletcher. They all probably made the alternative claim that some increase in performance could be obtained for the same energy consumption. However, – again probably – they tended not to quantify this. It was probably, indeed, that in reality it was not very great! They were all trying to put some of the dissipation losses of the main series resistances notches into car performance.

From the very first of the Second Phase of regenerative braking, field-control was necessarily incorporated in the modified controllers. This gave three notches to full series field function with Series 4 & 5 as field-weakening notches followed by parallel 7 to 11 notches incorporating further field weakening provision – the flanking interest mentioned earlier![10]

In pure series operation – that is, series motor – the field excitation in field control is provided by the current through the armature.

REFERENCES: APPENDIX 3

1. Markham, J.D., personal communication: 29 9 03
2. Brooks, R.: Electric Traction Handbook (Control): 1954: chII:13-14
3. Starr, A.T.: Generation, Transmission & Utilisation of Electric Power: 1949 – 2nd Edition: XII: 373
4. Ibid (Brooks): IV: 37
5. Ibid (Markham): 1 10 03
6. Ibid (Brooks): p16
7. Ibid (Brooks): 20
8. Electric Department: Coplawhill Car Works, Glasgow: Departmental Records
9. Ibid (Markham): 9 10 03
10. Fletcher, G.H.: Tramway Regeneration: Electric Railway, Bus & Tram Journal: 1931 – Sept. 18th: 134
11. Ibid (Markham): 7 9 02

APPENDIX 4

MALEY & TAUNTON – FIELD CONTROL

In Chapter 12 reference was made to the close similarity between the Johnson-Lundell (1904) scheme and the ideas promoted by Arthur Maley of Maley & Taunton in the mid-1930s.

Initially this was done under the banner of 'resistance-less control'. It utilised the same principle of having motor field windings in several sections, and changing their inter-connections through series, series-parallel, and parallel combinations. In effect transitions, but of the motor field windings only.

This is just what Lang's 1904 scheme did during motoring, but in fairness to Maley it could well have been a re-invention on his part. He might have got his inspiration from something as simple as the three heat rotary control switches (HIGH MED LOW) on electric stoves popular in the 1930s, which worked on an element combination selection basis – the similarity to Johnson-Lundell being quite coincidental. Maley could not copy the three heat switch circuit as its simplicity causes current reversal in one of the elements. This was immaterial for a heating circuit, but counter-productive for magnetic excitation.

Maley would be aware of remote control equipment for tramcars being promoted by others even though the 1927 BT-H contactor equipment on the prototype Metropolitan car 'Bluebell' suffered a disastrous runaway, resulting in the death of the motorman. Their 1929 equipment on the Swansea & Mumbles line was working well, and Met.-Vick Were promoting their electro-pneumatic system, one set having been installed on a London car.

Maley opted for electro-magnetic contactors, some of which were of the change-over pattern, with line voltage operating coils.

A big con, really?

The first in-service trial of Maley's 'resistance-less' system was probably on Birmingham 820, in 1934, as described in Chapter 5. It is very likely that the circuits were similar to, if not identical with, those used later on Glasgow 698 and 1100. Others in the traction manufacturing industry realised what Maley was up to, and JC Turrell, a traction control engineer with G.EC at Witton at the time, told me during the 1960s, after he had moved to Trafford Park, "It was a big con. really. What he had done was to subsume the resistances into the field windings, and put the heat where you don't want it – in the motors".

Considering the resistance values of the sections of field windings that are known for Glasgow, it appears that there could be truth in Mr Turrell's remarks. In their series combination with both motors also in series, the total field resistance was approximately 8 Ohms – a value close to what is required for a first notch standing start in a conventional tram with a resistance box under the stairs. The only difference of substance was that dissipation of the heat would be assisted by the fan on the armature drawing cooling air through the motor. There must have been a decrease in motor efficiency as a side effect, and overall such a system would give less by way of benefit in terms of total power consumed by the tramcar than first thoughts would imply. This was a result remarkably similar to that found in Newcastle in 1902, though where a different form of field control was in use (a compound winding with control of the shunt portion).

The reader is commended to the circuit diagrams of the Maley system as applied in Glasgow.

Birmingham remained unconvinced:

Having got about two years of service running in Birmingham with 820 'under his belt', Maley obtained Birmingham's rather reluctant permission to convert it to regenerative braking. This he did by an add-on process to his existing equipment, but in a way which marked the chief difference from the Johnson-Lundell system. The manual switch-over from motoring to regeneration effected by the motorman 'pressing a button on his controller handle' was not followed. Instead the regrouping of the motor fields was done when the controller was in the OFF position, and the command for regenerative brake initiation was taken from the handle of the air brake (Maley & Taunton, of course). By so doing the motorman would operate the controller in exactly the same way as in a conventional tramcar. No back-notching of the controller for regenerative brake (always necessary on the other systems up to that time) was required.

Any movement of the power controller from OFF dropped out the regenerative brake and allowed the motorman the use of the air brake. He could, therefore, operate power-coast points in the standard way.

Despite the success of Met-Vick's 50 Johannesburg regenerative cars (built in Birmingham by Met-Cam.), Birmingham Corporation remained unconvinced, perhaps with memories of the 1906 BMT problems, and the Yardley disintegration, not fully dispelled. Car 820 had the Maley electrical equipment removed in December 1937. It then reverted to Birmingham's standard. Could it be that some, or all, of 820's equipment later appeared in Glasgow?

Perhaps; but not the regenerative add-on.

Tramcar 'Overdrive'

The use of non-regenerative field-controlled cars was, apart from the Maley 'resistance-less' scheme, usually a way of increasing the maximum service speed which could be achieved on the level, and giving additional running notches. A number of systems tried odd cars (Birmingham, Glasgow) and a few (Metropolitan 'Feltham' cars, Blackpool and Liverpool) bought new fleets with this feature. The Liverpool 'Liners' all had both series and parallel weak field notches when new but it was certainly removed from all cars prior to the introduction of Automatic Emergency Braking (AEB) after the war. One of the problems suffered by these cars was damaged interpole coils due to overheating in the motors. Blackpool disabled all their's but then, it is thought, reintroduced it on to the ten English Electric Railcoaches converted to haul trailers.

If 'series' and 'parallel' are considered to be 'gears' in the automotive sense, the 'weak field' corresponds to 'overdrive'.

Weakening the field reduces the magnetising effect of the motor current on the field magnetic circuit. It reduces the tractive effort produced per ampere of armature current, and the Back-emf generated per mph The motor thus takes an incremental increase in current of a value often greater than that when going up one resistance notch. However, this typically produces a similar change in tractive effort due to the lower tractive effort per amp. The car accelerates to a higher balancing speed. This higher balancing speed is usually at a higher motor current than was the case at full field, and so raises the power demand on the supply system.

Leeds 272:

Jim Soper's book *Leeds Transport" Volume 3*, refers to car 272 being fitted with Maley field control. This may well have been the 'resistance-less' control similar to that on Birmingham 820 and the two Glasgow cars. It is unclear how this would be applied to BT-H 509 motors, although Maley – as far as is known - took motors free-issued by the customer, and had their field systems rewound with multiple coils to suit his scheme. If the BT-H motors in question were new, it is expected that BT-H would have none of this and would immediately withdraw any form of warranty concerning them. It would seem possible that the 'Maley' motors for 272 were older, warranty expired, machines that had been removed from another car. The 509s may only have been fitted after the Maley trial came to an end.

In correspondence with Mr Soper, John Markham records the former's strong evidence that BT-H 509 motors were indeed used, and continues by countering that:-

> "I have reason to believe that some versions of this family of motors were susceptible to commutation troubles. I know of cases in the UK(Leeds 180 at Crich), New Zealand and Australia, all of which point to this sort of problem.
>
> If these motors were at their limit in respect of commutation when working on full field (100%), then it is possible that they would react adversely with weak field operation. This would be aggravated even more by the very wide range of field control that Maley was trying to achieve, some of which, in the early notches in particular, may have been in excess of 100% when compared with conventional series operation. This could well drive parts of the field system into local saturation patterns such that the main field flux distribution became far from the optimum.
>
> I strongly suspect that there would be significant, to severe, to very severe, sparking at the brushes when operating under field control at anything other than quite low speeds. I also suspect that there would be an additional, more intense, flare-up each time a notch was taken.
>
> I am not a traction motor designer and therefore would find it very tedious to try to analyse from a theoretical point of view, just why these motors were as they were. This is not encouraged by the analysis I attempted in respect of Leeds 180 in the 1970s where the machine in question did not behave as simple theory said it should. The electrical neutral point was about 1½ bars out of position from the geometric neutral. I still do not understand why. The brushgear on the motor was moved to compensate for this.
>
> I fear Maley may well have faced similar problems with these machines which could possibly have been exacerbated by the neutral point moving as the percentage field strength varied.
>
> It certainly comes as no surprise to me that the trials of 272 with Maley field control did not last long!"

Leeds 255:

This car has already featured in Chapter 5 and was Britain's only four motor regenerative braking tramcar. Some fascinating additional detail comes to light from correspondence between co-authors that would otherwise probably never have become available.

Jim Soper's Leeds Book, Volume 3, states that Leeds 255 did have direct control comprising OK-type controllers initially, for about four months. The shunt field controllers would have been electro-magnetic (as they were on Glasgow's 305) and they remained as such even when the OK controllers were replaced by electro-pneumatic contactors. There is no evidence to even suspect that 255 ever operated before 'regen' and that 'regen' of one form or another was an original feature.

The wiring diagram (Leeds drawing 4332) shows the arrangement of the regenerative equipment and can only relate to 255 while in its electro-pneumatic regenerative state. This drawing is of interest in that the motor cut-out function is combined with that of the reverser, and not as on later equipments including the Middleton cars in Leeds, Liverpool and Glasgow cars, with the power brake change-over function. Leeds drawing 4691 relates to the production batch of Middleton bogies, and also 255 after conversion from regen.

As was seen in Chapter 5, some of Jim Soper's suggestions concerning the 'de-regen-ing' of

255 did not sit easily with the ex-Metropolitan-Vickers technical information in John Markham's possession, and none of the illustrations of the car Mr Soper uses is of the M-V pictures in this book, so that these two books will complement each other in these respects.

Back to Glasgow:

Reverting to the Maley & Taunton field control system, this was given a very full and extensive trial in Glasgow, on several cars, of which a summary of history is given in tabular form.

Maley & Taunton's senior electrical engineer in charge of the experiment was a small stocky little man, very clever and enthusiastic, of intense vigour and full of interest in the experiment. He spent immense time in the Coplawhill Car Works modifying and improving the electrics of the system. He was very much respected by the men who naturally nicknamed him "Snowey" on account of his white hair. Upon the closure of the Glasgow Corporation Tramways a number of technical drawings, much in disrepair and dirty, were salvaged and re-traced by the author for preservation purposes and these are also presented here.

Drawing 4741 is of the motor field circuits as applied to car 698, while 4678 shows the contactor panel diagram for the same car. The third drawing – 4403 – outlines the details of the contactor coils and resistances, probably for all three cars involved in the protracted experiment. Car 1100 is covered by Drawing 4651 for its controller and contactor panel diagram, while 4742 shows 1100's controller development. There were torn portions of other drawings, incomplete and dirty, and it is probable that there were many more that were not traced. These do represent some of the very many modifications and adjustments to the system of which the whole picture was simply not available. What can be presented, however, should go some way towards a general understanding of the complex principle of the field control system. It was not a failure, and functioned adequately under constant supervision, attention and modification: not the picture of economical function required for the rugged duties of city tramway service.

The M&T Field Control Saga:

The Initial Official Appreciation of the Maley & Taunton Field Control Testing:

In the disposal of office records when Glasgow's Head Office at 46 Bath Street was being closed, certain papers of tramway interest were salvaged. These included a batch on the Maley & Taunton Weak Field Tests, in their initial state, on Car 698. It should be noticed that J Gardner, Superintendent of the Motor School, referred to the system as M & T Electrical and Air Brake Control, and G Whyte, Motor Inspector 81, and Motor Inspector Larkin, who themselves tested Car 698 on Friday 25th August 1939, refer to it as simply the 'M&T Equipment'. In actual fact it was the recently fitted Field Control – or 'Resistance-less' equipment of Messrs Maley & Taunton.

All who worked on the early function of the equipment did so enthusiastically and John Markham, upon reading the letters agreed that one of them, at least, should be printed as the information was 'wonderful' and required incorporation into this section dealing with Car 698. It was of singular importance that Mr Maley's apparent success should be recorded. These interesting letters will be gifted by the authors to the National Tramway Museum for necessary preservation in their library, after publication.

Of the three reports, all were remarkably favourable. That of J Gardner is the least over enthusiastic and might possibly be the more reliable. However, Gardner only had a few runs from the Works up and down Albert Drive, in the presence of Mr Martin ('Snowey' the M&T Representative already referred to) along with David Shaw, then foreman electrical engineer, Coplawhill Works. The other Motor Inspectors Messrs Whyte and Larkin, were given the pleasure of taking the car out via Barrhead, Paisley and Govan by themselves with permission to put the car literally 'through the hoop'.

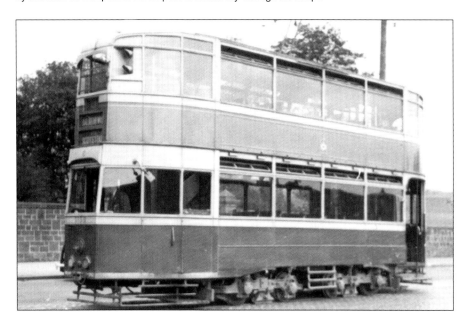

Glasgow 1100 was used as a test-bed for a number of experimental features, including field control.
(STTS collection)

"We committed every known fault…"

The three letters and reports are reproduced here:

Motor School

27th August, 1939

General Manager,

Sir,

Car No. 698. Maley-Taunton Electrical
and Air Brake Control.

On Thursday evening Mr. Dougan informed me by telephone that this car would be available for trial by any traffic officials I could muster for Friday morning.
I arranged for the two Motor-Inspectors procurable (White & Larkin) to attend at nine oclock on Friday morning.
With them I met Mr Martin, Maley-Taunton representative, who explained the equipment and answered some questions that occurred to me.
Although my other duties were tying me pretty closely at the time, I had trial and running demonstrations of the car in Albert Drive.
I wished the Motor-Inspectors to have something more than this and arranged for them to have full opportunity in a trial and practice run to Barrhead and back to the Car Works.
I saw them on their return and got the verbal report that from the driving and running point of view this equipment was the best that had yet come into their experience.
My own impressions were entirely favourable:-

(1)   The controller is the easiest imaginable to handle and is practically noiseless in operation.
(2)   Starting is perfectly smooth, smooth beyond comparison with the starting possible with any of the ordinary electrical equipment now or hitherto in use in Glasgow.
(3)   Acceleration is fast and perfectly free from any jerking.
(4)   The speed of the car is appreciably high and, although there was no speedometer, I was satisfied the car would comply with any requirements in this respect.
(5)   The driver's valve is of this firm's famous type giving such easy and wide selection of pressure and release as would afford drivers the maximum sense of security.
(6)   I was interested to know that the only alteration required to motors were in the windings and connections of field coils.
(7)   I gathered that losses through resistances were negligible.
(8)   I thought it a highly favourable feature, compared to the Department's remote control cars, that both power and magnetic brake are available in spite of any partial or complete failure of the air system. On remarking this to Mr. Martin and also that I had heard they had had some trouble owing to failure (occasional) of contactor operating coils, he told me they had not indeed had much trouble of this kind but they were considering the possibility of having to use air for the operation of contactors.
(9)   Mr Martin showed me some impressive figures of current consumption. My recollection is that the saving under this head was 35 to 40%.

The General Manager will probably recollect that the performance of this car in its so desirable qualities of smooth driving is in keeping with what I found on my visit to Edinburgh Transport Department on the 27th of December, 1934 and which I duly reported on my return.

Yours truly,
(signed)   J. GARDNER.

GLASGOW CORPORATION TRANSPORT
ROAD TEST OF MALEY-TAUNTON EQUIPMENT

Motor Inspector W.G. Whyte.   Friday, 25th August, 1939
   Car No. 698   9 a.m.-12.30 p.m.

Sir,
In presence of Mr. Martin (Maley-Taunton representative) and Mr. Shaw (foreman electrical engineer from Coplawhill Works) I tested above car from 9 a.m. till 12.30 p.m.
  Travelling from Works we ran car up and down Albert Drive under the supervision of Mr Gardner (Supt Motor School) and tested power of acceleration and braking, both air and magnetic. Car was then taken to Barrhead and severely tested on private tram track before returning to Works, via Govan and Shields Road.
  We were permitted to do as we liked with equipment, and I can assure you, Sir, we committed every known fault, in the endeavour to find any weakness in the electrical equipment, but despite the abuse and mishandling received, equipment overcame all tests. I am positive in saying that no other type of equipment, at present in use in Department, would have withstood these rigorous tests without extensive damage.
  The points that please best and make this type an improvement on **all** other equipment at present in use are:-

(1) Ease of operation and smoothness of acceleration
(2) Electro-magnetic brake available, even at lowest speed, and not dependent on air supply, as in remote control Met-Vick System.
(3) Smooth action of magnetic brake when applied.
(4) Absence of noise in units when operating same.
(5) Ease of 'motors cut outs' and access to 6 amp. Unit control fuse.
(6) General speed.
(7) Ease of conversion to this equipment from standard type.

The only thing against this type we could find is that the '4th method' is not available.

Yours resp
(Signed)   W.G. WHYTE
         M/Inspr. 81

GLASGOW CORPORATION TRANSPORT

ROAD TEST OF MALEY-TAUNTON EQUIPMENT

Motor Inspector Larkin.   Friday, 25th August, 1939

Sir,

Regarding car No. 698 fitted with Maley Taunton equipment.

I tested the car and found the acceleration very good, Brakes most effective and controls simple to operate.

In my opinion the car gave every satisfaction.

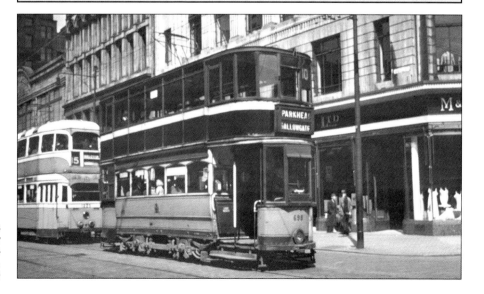

Glasgow Standard car 698 was handed over to officials with a view to their putting it through its paces to test the Maley & Taunton equipment. Here it is in 1951. *(Courtesy C Carter)*

INDEX

TAILPIECE

Having shown full-page illustrations of modern trams using regen braking in France and Germany it seemed churlish not to include Switzerland in what its Editor has dryly described to me as "this book about *The Regeneration Game*".

The view below of the tramway station outside the Hauptbahnof in Basel will remedy the situation. An immaculate Schindler 5-car set, led by number 214, from the yellow and red-liveried BLT system (on the left) loads alongside one of the latest Siemens Combino cars, number 327, of the city system in the new lighter green livery in this sunny October 2006 morning scene. Travel on either is a wonderful experience and a salutary reminder of what most commuters in the UK are missing – fast, reliable, comfortable and smooth running ecologically-friendly public transport.   *(Photo: John A Senior)*